# The Cultural Politics of Digital User Experience Writing

User experience (UX) writers are the professionals who create the verbal content of websites, apps, or other software interfaces, including error messages, help texts, software instructions, or button labels that we all see and engage with every day. This invisible yet highly influential language work has been largely ignored by sociocultural linguists. The book addresses this gap, examining the broader cultural politics of digital media through an exploration of the linguistic production and purposeful design of interface texts. It discusses UX writing as an influential contemporary domain of language work and shows how the specific practices and processes that structure this work shape the norms that become embedded in software interfaces. It highlights the nature of UX writing, its (meta)pragmatic organization, and its cultural-political implications. Foregrounding the voices and perspectives of language workers, it is essential reading for anyone interested in how language shapes the way people use digital media.

LARA PORTMANN holds a PhD in Language and Communication from the University of Bern, Switzerland. She has also worked in user experience design for nearly a decade. Through her research, she strives to build a bridge between academic research and industry practice.

# The Cultural Politics of Digital User Experience Writing

Lara Portmann
*University of Bern*

CAMBRIDGE
UNIVERSITY PRESS

CAMBRIDGE UNIVERSITY PRESS

Shaftesbury Road, Cambridge CB2 8EA, United Kingdom

One Liberty Plaza, 20th Floor, New York, NY 10006, USA

477 Williamstown Road, Port Melbourne, VIC 3207, Australia

314–321, 3rd Floor, Plot 3, Splendor Forum, Jasola District Centre, New Delhi – 110025, India

103 Penang Road, #05-06/07, Visioncrest Commercial, Singapore 238467

Cambridge University Press is part of Cambridge University Press & Assessment, a department of the University of Cambridge.

We share the University's mission to contribute to society through the pursuit of education, learning and research at the highest international levels of excellence.

www.cambridge.org
Information on this title: www.cambridge.org/9781009540582

DOI: 10.1017/9781009540605



When citing this work, please include a reference to the DOI 10.1017/9781009540605

First published 2025

Cover image: Filo/Digital Vision/Getty Images

*A catalogue record for this publication is available from the British Library*

*Library of Congress Cataloging-in-Publication*
Names: Portmann, Lara author
Title: The cultural politics of digital user experience writing / Lara Portmann.
Description: Cambridge, United Kingdom ; New York, NY : Cambridge University Press, 2025. | Includes bibliographical references and index. | Summary: "Drawing on a range of interviews with digital user experience (UX) writers, this book offers a thorough introduction to how language is used in the production of websites, apps, and other software. It is essential reading for anyone interested in how language shapes the way people can use digital media"– Provided by publisher.
Identifiers: LCCN 2025003375 (print) | LCCN 2025003376 (ebook) | ISBN 9781009540582 hardback | ISBN 9781009540612 paperback | ISBN 9781009540605 epub
Subjects: LCSH: User interfaces (Computer systems) | Technical writing–Social aspects | Online authorship–Social aspects | Cultural pluralism
Classification: LCC QA76.9.U83 P65 2025 (print) | LCC QA76.9.U83 (ebook) | DDC 744.8/5–dc23/eng/20250411
LC record available at https://lccn.loc.gov/2025003375
LC ebook record available at https://lccn.loc.gov/2025003376

ISBN 978-1-009-54058-2 Hardback

# Contents

# Figures

# Acknowledgements

Writing a book is hard. It is an arduous and often solitary process, which is why I am all the more grateful to the people who have helped me along the way.

This book is based on my PhD research. I am forever grateful to my thesis supervisor, Professor Crispin Thurlow, who believed in this project from the very first time we sat down to discuss it. Thank you for your enthusiasm for my research, for your intellectual guidance, and for your continuing support. Working as your PhD student has been an honour and a privilege.

In this respect, I am also thankful to Professor Rodney Jones, who acted as the external examiner of my dissertation and whose helpful and generous feedback on my research guided the revisions of this book.

My research was conducted within the framework of a research project funded by the Swiss National Science Foundation and led by Crispin Thurlow, titled *Elite Creativities: Engaging the Language Work of Professional Wordsmiths* (100015_192299). I am grateful to the other members of the *Elite Creativities* team, Gwynne Mapes and Olivia Droz-dit-Busset, for our many exchanges – research-related as well as otherwise. Thank you, also, to all my other friends and colleagues at the University of Bern.

Thank you to the editorial team at Cambridge University Press, and to Rebecca Taylor in particular for her expert guidance and continued helpfulness. I am also grateful to the anonymous reviewers of this book for engaging with my work and providing such valuable input. Thank you, also, to Isabel Collins and Natasha Whelan at Cambridge University Press and to Dhanujha Harikrishna at Straive for their kind and patient support with the final stages of this book.

Thank you, especially, to all the amazing UX writers I met while conducting this research. I knew that the UX writing community was welcoming, friendly, and open-minded, but I did not expect that my interviewees would be so generous in sharing their work, their opinions, and their expertise with me. My thanks to all of you; this research would not have been possible without you. Special thanks go also to Fiona Tinner and Tamara Johnson, with whom

I started organizing the UX Writing and Content Design Meetup (Bern & Zurich) while writing this book.

I want to give a particular shout-out also to Angela Meyer and Jasmina Slisko, who asked me to send them my chapters just because, and who kept cheering me on with their enthusiasm and interest in my work.

Finally, I wish to thank my family, especially the two zoms. You have kept me safe and sane during this journey, from the beginning to the very end. There are no words.

# Introduction

# 1 The Production, Pragmatics, and Politics of Digital Media

Language is at the heart of digital media, although sometimes in obscure and not necessarily obvious, visible ways. In fact, language in digital media interfaces is often specifically designed in such a way that it (ideally) goes unnoticed. As a whole, my book is concerned with the cultural politics of digital media, and the way digital media interfaces assert and reproduce existing, normative discourses. I examine how digital media interfaces are written or *designed* – that is, purposefully crafted – and how such design choices are enmeshed with social, economic, and (cultural-)political matters. Specifically, I focus on the working practices of so-called user experience (UX) writers. Typically working behind the scenes and often invisibilized, these are the professionals who create the verbal content of websites, apps, or other software interfaces.

According to UX writer Lisa Sanchez, people encounter UX writing whenever they use the internet, but 'when it's doing its job well, you don't even notice it' (para. 8).[1] As such, UX writing influences what people can do with digital media. Often, this occurs in banal, seemingly harmless but actually consequential ways. As a first example, I offer Figure 1.1, which shows the sign-up interface of Facebook in 2022. Inconspicuous as this interface may be, it makes a number of assumptions about who can and who should be using Facebook. Clearly, there is a default or ideal user that is inscribed in this text: someone with a first and a last name, with a known date of birth, and who is preferably female or male. While the interface also allows users to enter any or no description of their gender by selecting the 'custom' option, the fact that 'female' and 'male' are presented as default options emphasizes a binary and heteronormative view of gender (cf. Bivens, 2017). I will return to these normative aspects later, but for now, my point is simple: the work of UX writers is both banal and at the same time influential; the texts they produce effectively shape (or even dictate) what people can do and who they can be when using digital media. Texts like the ones in Figure 1.1 are everywhere in

[1] Source: *What is UX Writing?* Retrieved from https://lmsanchez.medium.com/what-is-ux-writing-1eb71b0f0606 (accessed 6 October 2022).

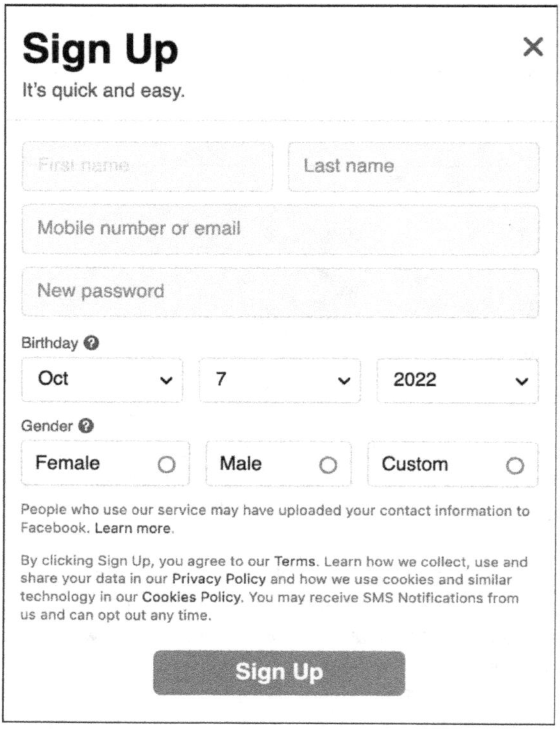


**Figure 1.1** Facebook's sign-up interface in 2022

people's interactions with digital media and yet, as citizens and as scholars, people often misrecognize their influence. Certainly, only little is known about the circumstances under which these texts are written and designed.

While scholars in digital discourse studies have examined in depth how people communicate with and through digital media (e.g. Thurlow & Mroczek, 2011; Jones et al., 2015; Georgakopoulou & Spilioti, 2016), only some scholars have taken up interface texts as objects for critique (e.g. Chik, 2015; Hafner, 2015; see also Djonov & van Leeuwen, 2017, on semiotic software). Regardless, there is usually little consideration given to the linguistic work that goes into producing interface texts. Indeed, academics and everyday users of digital media alike may not even be aware of the specific professional roles that shape the texts they see when they interact with software.

It is precisely the linguistic production of interface texts that I seek to address in this book. I do so by looking not just – or not even primarily – at these interface texts themselves, but by foregrounding instead the language work that goes into producing them. Concretely, I am interested in UX writing as an influential contemporary domain of language work, and in how the specific practices and processes that structure this work shape the norms that become embedded in software interfaces. Grounded in sociocultural linguistics, my project thus orients to two major research traditions: *digital discourse studies* (specifically with regards to recent work addressing the cultural politics of technology and so-called digital pragmatics, e.g. Djonov & van Leeuwen, 2017; Jones, 2020b), as well as research in *critical sociolinguistics* on language work/ers (e.g. Duchêne & Heller, 2012; Thurlow, 2020b). Before elaborating on this broader theoretical framing of my research, however, I offer a brief introduction to UX writing itself.

## 1.1 UX (User Experience) Writing: A Brief Introduction

Simply put, UX writers are contemporary language workers typically employed in web and software design companies, where they are responsible for writing the words ordinary users see whenever they interact with digital media. This might include the text on buttons that users click, error messages they see, or the labels of online forms (recall Figure 1.1). UX writers typically refer to these little texts as *microcopy* or *(user) interface copy*. The day-to-day work of UX writers is of course more complex than this; for now, however, suffice it to say that microcopy is the most emblematic linguistic output UX writers produce as part of their work.

Microcopy itself is evidently not a new phenomenon. Indeed, before the emergence of the graphical user interface (GUI), digital media were almost entirely text-based (e.g. Bolter & Grusin, 1999, pp. 23, 197). In light of this, it is all the more interesting that the job title 'UX writer' only surfaced in the mid-2010s, and that the task of writing the words which appear in user interfaces thus only recently received attention as 'proper' work within the software design industry. To be clear, I am not suggesting that the kind of work UX writers do is novel or unprecedented; as mentioned, words have always been part of software interfaces. Still, what makes this profession compelling for scholarly investigation is the fact that many UX writers themselves claim that this work is new, and that they put considerable effort into establishing UX writing as its own professional field, all of which influences their work, including the interface texts they produce as part of that work.

It is worth noting that UX writing is also a relatively prestigious profession. The UX writers I encountered through my research were all proud of their work and expressed strong personal identification with their jobs: the title 'UX writer' appears to have a particular kind of cachet in the broader professional field that UX writers are situated in; certainly, it bears none of the stigma associated with other kinds of language work, such as that of call centre workers (cf. Woydack, 2017). As part of the so-called tech industry (cf. Dorschel, 2022), UX writers also tend to earn a relatively high salary. A recent industry survey reports, for instance, that the average salary of a UX writer in the US in 2019 was USD $126,000; the average salary in the US in 2019 was USD $66,400.[2] The fact that the work of UX writers is so highly valued – both in symbolic and material terms – is in part what makes them interesting to me as elite language workers or *wordsmiths* (Thurlow, 2020c, more on this term below). As outlined above, UX writers are also quite influential language workers, as they – together with others working in software design – shape the use and possibilities of contemporary digital media. But their work is typically done behind the scenes, and so its influence or impact is often misrecognized and obscured. It is exactly this interplay between high-status language work and cultural-political impact that interests me here. This leads me to the broader theoretical framing of the current book: the cultural politics of digital media and the economy of language work.

## 1.2 Digital Media as Cultural Politics

Situated within digital discourse studies, this book offers a sociolinguistic investigation of the role of language in the design of software interfaces and the (cultural-)political consequences of such design choices. As such, the book addresses how language is entangled in particular, normative interactions users have with and through digital media (cf. Jones, 2020a). By 'cultural politics', I mean the way culture (including people's attitudes, opinions, beliefs, and perspectives) comes to shape society and give rise to particular social, economic, and legal realities (cf. Nash, 2001). In speaking of the cultural politics of digital media, I want to draw attention to the fact that technological artefacts are not neutral, neither in their design nor in their use. Instead, they are bound up with specific ways of being and doing things in the world; they are inscribed with and in turn re-inscribe particular norms and identities.

Arguing that digital media are political is, in some ways, unremarkable. In an era when politicians use social media platforms like Twitter (Now X) as major publicity channels (e.g. Lee & Shin, 2014; Frame & Brachotte, 2015), it has become clear that the technologies people use to socialize can and are also deployed to gain political advantage. In this way, one could say that

[2] Sources: *UX writer salary survey: How much money do UX writers make?* Retrieved from https://uxwriterscollective.com/how-much-do-ux-writers-make-salary-survey-revealed (accessed 10 October 2022); *Average annual real wages in the United States from 2000 to 2020.* Retrieved from www.statista.com/statistics/612519/average-annual-real-wages-united-states/ (accessed 10 October 2022).

'technologies are political in that they can be used for political ends' (Slack & Wise, 2014, p. 165). While this is certainly true, it is, as Slack and Wise suggest, an oversimplification. Such a view sees technology as a tool, politics as the dealings of political parties, and understands the relation between both as politics being 'done' via technology. However, if one views technology and politics from a cultural point of view, a more complex picture emerges.

From a cultural perspective, politics can refer not just to political parties, but also to how relations of power are configured, exercised, and challenged in all aspects of social and cultural life (cf. Kramsch, 2021); technology then becomes 'an assemblage in which political and economic work is performed' (Slack & Wise, 2014, p. 168). Such an approach rejects seeing technology as a neutral tool. Instead, culture and technology are understood as inherently enmeshed, with one constituting the other and vice versa. In short, there is no culture *and* technology but *technological culture* (Slack & Wise, 2014) or *technoculture* (Ross, 1990; see also Roderick, 2016), which is both material and semiotic at the same time. It is this view that I take when speaking of the cultural politics of digital media.

### *1.2.1 The Cultural Politics of Technology*

Cultural-political approaches to technology are not new. A prominent example of such scholarship is Winner (1980), who in his 1980 essay of the same name asks: 'Do artefacts have politics?' Critiquing both technological determinism and the social determination of technology, Winner argues that technologies 'can embody specific forms of power and authority' (p. 121). For Winner, there are two ways in which technological artefacts may do this. The first are instances where the design or arrangement of an artefact has a political effect. Winner's famous (and sometimes contested) example is the case of the low-hanging highway overpasses on Long Island, New York, designed by Robert Moses.[3] Because of their design, the highway overpasses made it impossible for buses to pass through, exerting a politics of racial and class-based exclusion: upper- and middle-class white people (i.e. those wealthy enough to own a car) could use the highway, whereas poor and typically Black people, dependent on public buses, could not. Consequently, Moses's design limited the access of low-income groups to his widely acclaimed public park, Jones Beach. For Winner, this narrative shows how artefacts are never just what they appear to be; rather, they are always also ways of building social order in the world.

[3] There is some debate around the veracity of Winner's account (e.g. Joerges, 1999). Nonetheless, Winner makes a compelling case for how artefacts are never just neutral 'things'; they always also have sociocultural impact.

Winner's second case of how technologies are political are what he calls 'inherently political technologies' (p. 123), which are systems that through their very nature require (and thus construct and reinforce) particular political circumstances. To choose these technologies entails subscribing to a particular form of political life. As an example Winner discusses large-scale technological systems such as oil pipelines, which, in order to work effectively, require centralized and hierarchical control. Here, it is the very nature of the technological system that necessitates a particular mode of power and authority. As such, Winner makes a compelling case for how artefacts are always bound up with the norms and values of those who construct and/or control them.

Like Winner's oil pipelines, digital media can also depend on large-scale systems that favour centralized and hierarchical modes of control. For instance, the seamless, rapid delivery systems of an e-commerce giant like Amazon are only possible because of the strict, hierarchical, and sometimes inhumane conditions to which its warehouse workers are subjected.[4] Similarly, food delivery apps such as Foodora or Deliveroo, often hyped as innovative and progressive technology start-ups, depend on centralized and non-transparent managerial control of human labour (Ivanova et al., 2018). Beyond such examples of the potentially 'inherent' political nature of digital media, digital media are also political in terms of how their designs build order in the world. In this regard, Djonov and van Leeuwen (2017) argue that software always comes with a built-in 'semiotic regime' (see also Poulsen et al., 2018). Using the example of PowerPoint, Djonov and van Leeuwen (2017) show that the software itself not only imposes norms about what counts as an effective presentation, but that it may also 'intervene in the ways people use language and other semiotic resources' (p. 579). In other words, software design shapes how people are able to make meaning in the world – or not.

Another obvious example of the politics of digital media are sign-up interfaces, which quite literally ask users to identify themselves in terms of pre-defined categories (recall Figure 1.1). This ordering of people into pre-defined social categories – usually based on a Western, heteronormative worldview – is often connected to commercial interests. A fashion retailer might ask people who sign up to their newsletter to indicate their gender in order to provide more 'relevant' news and discounts. Typically, this means showing users ads and incentives that are more likely to incite them to buy

[4] Source: *Amazon workers working 55-hour weeks and so exhausted by targets they 'fall asleep standing up'*. Retrieved from www.independent.co.uk/news/uk/home-news/amazon-workers-working-hours-weeks-conditions-targets-online-shopping-delivery-a8079111.html (accessed 8 September 2022).

something. Others yet may sell such data off to third-party advertisers, who in turn can show users more 'relevant' advertisements. These commercial logics can become deeply intertwined with the design of an interface. For instance, so-called cookie consent notices are typically written and designed in such a way that they leave users almost no other option than to accept cookies (Jones, 2020a; see also Portmann, 2022). Another poignant example of how commercial interests shape digital media is Facebook, which until mid-2019 used to ask new users to select 'male' or 'female' when signing up – long after the options for gender identification that the platform offered to already registered users had been redesigned. While users were free to change their settings once logged in, Facebook asked for an initial binary identification in order to retain the commercial advantage of being able to classify users into traditional, gendered marketing categories (Wachter-Boettcher, 2017). Once again, normativity is here directly linked to commercial interests.

In the case of Facebook, the underlying commercial and computational logic is so strong that even today, with the possibility to choose a so-called 'custom' gender option when creating a new Facebook account, beneath the surface, custom gender options are still re-categorized into a heteronormatively coded database (Bivens, 2017). Thus, while the user-facing side of Facebook allows people to choose out of more than fifty gender options, in its underlying database, the platform classifies people into the categories male, female, and custom.[5] What is more, Bivens (2017) shows that Facebook's code determines these values not by evaluating what users enter into the 'gender' field, but through the pronouns they pick. Following this computational logic, even someone choosing 'gender questioning' will be re-coded as 'female' if that person also selects the pronoun option 'she/her'. In the end, the platform's custom gender project is therefore bound up with binary gender norms, which remain aligned with the company's financially motivated goals.

It is tempting to dismiss ideological decisions in software and interface design on the grounds of technical limitations. The digital is by definition discrete, and computer code is by its very nature binary: this might be another way in which digital technologies are (or at least seem to be) 'inherently political' (Winner, 1980, p. 123) – implementing software that accounts for a fluid conceptualization of gender may simply be incompatible with the binary nature of code. However, it is important to consider how such (apparent) limitations are treated. For instance, despite the binary and discrete structure of code, more fluid representations of gender are indeed possible, as a handful of video games which employ a 'gender slider' instead of 'either/or' options

[5] In some legacy cases, users may also be classified as 'undefined' (Bivens, 2017).

demonstrate.[6] Decisions about whether (and which) technical limitations are addressed can therefore be understood as an indication of which problems are deemed worth solving.

An example in Benjamin's (2019) discussion of race and technology nicely exemplifies the unequal ways in which technical limitations tend to be addressed or not. She recounts that, in 2013, media specialist Allison Bland tweeted that the voice directions of Google Maps had instructed her to 'turn right on Malcolm Ten Boulevard'. Google Maps did not recognize the name of Black civil rights activist Malcolm X, instead interpreting the 'X' as the roman numeral ten. As Benjamin points out, this decoding of 'X' as 'ten' is a case of misrepresentation; ironically, however, having a computer correctly parse 'X' as 'ten' is, from a technical standpoint, actually a significant achievement. This tension between technical achievement and social misrecognition is exactly the point, Benjamin argues: 'This illustrates how innovations reflect the priorities and concerns of those who frame the problems to be solved, and how such solutions may reinforce forms of social dismissal' (p. 84). Effectively, this is the digital equivalent of Moses's highway overpasses. While Google Maps does not literally prohibit Black people from going anywhere, this glitch, which is perhaps 'better understood as a form of displacement or digital gentrification' (Benjamin, 2019, p. 83), also tells people where they do and do not belong. Such matters are clearly rooted in issues of production and connected to who designs digital media. It is no secret that the software design industry is disproportionately white, middle-class, and male; this structural privilege inevitably impacts the software that is produced (Bettivia, 2016). Although frequently veiled in claims about neutrality and objectivity, computers and code are also implicated in the articulation and reproduction of racist, sexist, and classist systems of belief (cf. Nakamura & Chow-White, 2012). Digital media, too, have politics.

### *1.2.2 The Politics of the Interface*

As the main way in which people access and interact with contemporary digital media, software interfaces are particularly key in how social norms are perpetuated through digital media. Interface, however, is a complex term. Even if one reduces the scope to digital technologies, 'interface' has multiple meanings. While most people may think of the graphical user interface as 'the' interface, user interfaces can also take the form of, for instance, voice or command-line interfaces. What is more, interfaces in the context of digital

[6] Source: *The gender slider – LGBT+ representation in gaming.* Retrieved from https://thenorwichradical.com/2016/06/11/the-gender-slider-lgbt-representation-in-gaming/ (accessed 8 October 2022).

media need not necessarily be *user* interfaces; there are, for instance, also application programming interfaces (APIs), which specify interactions between different software parts. That said, in this book, I use the term interface to denote user interfaces, and most of the examples I discuss are examples of graphical user interfaces – the kinds of interfaces UX writers typically work on.

By framing the choices people have when interacting with digital media, user interfaces both enable and constrain users. They allow people to use digital media without having much knowledge about the underlying technological system, yet they also limit them to the options the interface makes visible. For Monteiro (2017), the interface thus 'remains an important site in contemporary culture—perhaps the most important site—for the function of power on, and through, the individual' (p. 8). As such, interfaces give insight into the politics of contemporary digital media, indicating how and by whom digital media are (supposedly) meant to be used.

Again, understanding software interfaces as political is not necessarily a new perspective. In 1994, rhetoric and media scholars Selfe and Selfe (1994) suggested that '[c]omputer interfaces ... are also sites within which the ideological and material legacies of racism, sexism, and colonialism are continuously written and re-written along with more positive cultural legacies' (p. 484). Three decades later, it seems that not much has changed. Today's interfaces are still structured by the same white and middle-class principles that Selfe and Selfe (1994) identify: the computer is still conceptualized as an 'office' with a desktop, folders, and files (a decidedly middle-class environment, as Selfe and Selfe point out). At the same time, English is still the dominant language of digital media. For instance, even as the web is getting more multilingual support, program shortcuts remain based on English: CTRL + S for '**s**ave', CTRL + C for '**c**opy', CTRL + P for '**p**rint', and so on.[7] In this regard, it is of course also telling that technical support for scripts other than the Roman alphabet are often deemed second priority, supposedly due to technical limitations. This is not a necessity, but something that has persisted out of convenience, all while reinforcing an anglophone and typically middle-class reality.

The 'legacies of racism, sexism, and colonialism' Selfe and Selfe (1994) speak of run so deep that it is only in light of the renewed public interest in the

[7] Even seemingly arbitrary shortcuts like CTRL + Z for 'undo' and CRTL + V for 'paste' are rooted in US English, having likely been chosen based on their relative location to other shortcut keys such as CTRL + C (mnemonically for 'copy') and CTRL + X (for 'cut', likely a visual mnemonic): on a US American keyboard layout, the letters Z, X, C, and V occur in a single row next to each other. On other keyboard layouts, the letter Z, for instance, often appears in a different row. Of course, for keyboard layouts based on other scripts, these shortcuts make even less sense.

Black Lives Matter civil rights movement in 2020 that some major software companies such as the Microsoft-owned GitHub finally started working on removing racist coding terminology such as 'master/slave' and 'whitelist/ blacklist' from their codebase.[8] While it is surely laborious and time consuming to replace these words in thousands of lines of code, the remedial actions of GitHub demonstrate how changes are not impossible. Like the Google Maps glitch mentioned earlier, such decisions are a matter of priority, where inaction is as telling as action (see Benjamin, 2019, discussed above). How digital media are written and designed thus remains a normative process. Additionally, what emerges is that the cultural politics of digital media depend neither on producers nor users alone, rather, producers, users as well as the software itself are part of a complex technocultural system. It is in this way that I turn also to posthumanist approaches in order to address the politics of digital media interfaces.

### *1.2.3 Posthumanist Approaches*

Posthumanism is an umbrella term used for a range of allied, contemporary movements in the humanities and humanistic social sciences, which, at their core, are concerned with questioning the notion of the 'human' (cf. Herbrechter et al., 2020). Concretely, posthumanism rejects the idea of the (rational) human as central or exceptional vis-à-vis non-human entities. This entails questioning the humanism of the Enlightenment, acknowledging that the humanist conceptualization of human (including the concomitant notion of human rights) is not inclusive but exclusive (e.g. Herbrechter, 2013; Pennycook, 2018a). As such, posthumanism aligns most obviously with the formative work of science and technology scholar Haraway (1991). While Haraway herself rejects the label posthumanist for her own work (Haraway & Goodeve, 1998), her *Cyborg Manifesto* – which argues for a breakdown between human and animal, human and machine, and physical and non-physical – crucially informs posthumanist thinking; her theorizing of gender equally addresses matters of who is considered 'more' and 'less human'.

Evidently, distinctions between 'more' and 'less' human are still prevalent today, and one way in which they surface is through technology. For instance, cars are typically manufactured according to standards which protect only male crash dummies (Criado Perez, 2019), automatic soap dispensers frequently respond only to white skin (Benjamin, 2019), and almost 95% of the most popular Android apps are said to be only partially accessible to people with disabilities (Yan & Ramachandran, 2019). Seemingly meant for everyone,

[8] Source: *GitHub abandons 'master' term to avoid slavery row*. Retrieved from www.bbc.com/news/technology-53050955 (accessed 8 October 2022).

these different kinds of technologies – cars, soap dispensers, and mobile apps – turn out to be made only for particular groups of people. In this regard, posthumanism acknowledges also that the human is not a singular notion but a plural one, understanding 'human' as a socially constructed category.

Lamb and Higgins (2020) note five key areas of posthumanist thought in the social sciences and humanities: (1) critiques of humanism, which address and complicate dominant understandings of humanism and the human as based on Enlightenment thinking; (2) actor-network theory (ANT), which is perhaps best known for stressing the agency of non-human entities and complicating the human as the sole source of action; (3) assemblage theory, which, similarly to ANT, stresses relational exchanges and relations between human and non-human entities, though emphasizing the fluidity and unpredictability of these relations; (4) new materialism, which emphasizes our entanglement with the material world and the active role of matter; and (5) critical plant and animal studies, which, as the name suggests, interrogate the often arbitrary boundaries that are being drawn between the human and other living organisms.

Acknowledging that the term posthuman encompasses all this, my current discussion draws most strongly on ANT (e.g. Latour, 2005) and new materialism (Barad, 2007) for understanding power and agency in the context of technoculture. ANT sees the social (and thus culture) as a series of connections, constituted by human and non-human participants (e.g. Callon & Latour, 1981; Latour, 2005). In this view, agency (which is not to be confused with intentionality), is not limited to the human but can also be held by non-human entities. If one accepts that the non-human can be an actor (which is to says that it can cause an effect in a sociotechnical system), separating the human from the non-human becomes pointless: both contribute to making up the social (cf. Latour, 1990). Similarly, Karen Barad's agential realism situates agency not in individual beings or things but sees it as 'a matter of intra-acting' (Barad, 2007, p. 178). Unlike the traditional notion of interaction (where separate entities act upon each other), the term intra-action emphasizes the fundamental entanglement of human and non-human entities, whereby subject and object are seen as always co-constitutive, emerging precisely through particular intra-actions. This split into subject and object is enacted through multiple, cross-cutting 'agential cuts' through which particular boundaries are rendered meaningful. Bearing this in mind, I am interested in understanding how the category of 'human' is constructed in such sociotechnical systems through material-discursive intra-actions and in how UX writers conduct their language work from within complex assemblages in which they are subject to specific configurations of agential cuts which lead to the emergence of particular professional identities and norms. In this way, I seek to address some of the normative processes involved in the production of digital media by focusing on the language work of UX writers.

This brings me to a second field in sociocultural linguistics I orient to: critical sociolinguistic research on language work.

## 1.3 The Economy of Language Work

In examining how language is taken up as a resource in the design of digital media, I orient to scholarship on language work. As a whole, language work has been linked to the shift from manufacturing-based economies to service-based economies and the ensuing commodification of knowledge and communication. In this vein, scholars in discourse studies and critical sociolinguistics have sought to address how language practices relate not only to society and culture but also to economy, arguing that language can be commodified and (re-)negotiated as a source of profit (cf. Duchêne & Heller, 2012). Scholars like Fairclough (1999), Cameron (2000b), and Heller (2003) have shown in depth how language and work have become increasingly entangled in the wake of late capitalism. Not only do professions that used to depend on skills and resources other than language now often require communicative skills, but language in itself is also increasingly the central focus of people's work. For businesses, language can become a potential source of competitive advantage as well as a factor in business productivity (Boutet & Heller, 2007), something that can be – or, it would seem, must be – actively managed (Boutet, 2012; see also Cameron, 2000b). For Heller and Boutet, this is what has led to the emergence of language work and the language worker – professions where 'the language part of work' (Boutet, 2012, p. 208) is not just part of people's work, but its main point.

Language work in late capitalism is often seen as enmeshed with more symbolic uses of language. In this way, linguistic practices in call centres may, for instance, take on stylized and at times highly regulated forms (see e.g. Cameron, 2000b). The importance of such symbolic aspects of language is something that Kelly-Holmes (2014, 2020) observes quite pointedly with regards to the use of multilingualism in marketing. Speaking of the 'linguistic fetish', she shows that in the linguistic business of marketing, languages are often used not for instrumental purposes but 'to enhance the semiotic landscape of the advertisement' (Kelly-Holmes, 2020, p. 38). Through its symbolic meaning, language can thus become a source of added value in the 'saturated markets' of late capitalism (cf. Heller, 2010b). This is how language becomes part of the production, circulation, and consumption of resources; in other words, language becomes a matter of economy.

Of course, not all language work is equal. While language scholars have examined at length the work of low-status language workers such as call centre workers, it is only recently that scholars have started paying attention to more elite language workers such as marketers, PR officers, or academics themselves (cf. Thurlow, 2020c). Following Thurlow (2020c), I understand the adjective 'elite' here not necessarily as a sociodemographic category, but as a relational status that is discursively constructed and materially rooted (cf. Thurlow & Jaworski, 2017a). These more elite language workers are what Thurlow (2020c) calls 'wordsmiths' in order to better map the relative hierarchies of language work. Wordsmiths are *relatively* speaking more privileged – for instance vis-à-vis other, low-status language workers. That privilege can be material (e.g. a higher salary) or symbolic (e.g. more influence); importantly, though, it is always also discursively negotiated, for instance through particular appeals to professionalism and/or expertise. As Thurlow (2020c) suggests, what makes wordsmiths powerful is that their work typically hinges on its behind-the-scenes nature while nonetheless being influential in people's everyday lives and potentially instrumental in shaping societal norms and social values. It is these two aspects – status and normativity – that sit at the heart of my book. However, before I turn to my research design, I briefly want to clarify my understanding and use of the term 'political economy' as it relates to language work.

In a nutshell, political economy is centrally concerned with the allocation of resources. In writing about the *Political Economy of Communication*, Mosco (1996) defines the term as describing 'the study of the social relations, particularly power relations, that mutually constitute the production, distribution and consumption of resources' (p. 25). More generally, political economy can be understood as a sociopolitical approach to economy based on the fundamental understanding that material economy cannot be separated from social foundations and power relations (cf. Stilwell, 2019); as such, political economy orients to and addresses matters of social class and/or social stratification.

Looking to research from language scholars, there seem to be two primary traditions in the way scholars tend to operationalize the term political economy. On the one hand, there are some who assert that scholarship done under this rubric ought to thoroughly engage 'with the past and present of political economy' (Block, 2017, p. 53), i.e. both with classic political economy from Smith to Marx and with contemporary work on current economic developments. Research in this tradition typically entails a strong orientation to Marxism and economics (e.g. McGill, 2013; Block, 2018; see also papers in Petrovic & Yazan, 2021b). On the other hand, there are scholars who take a less explicitly economic approach, and who are more interested in using the term political economy as a way of highlighting the material (and thus necessarily also economic) nature of language, aiming to address how language is always tied to a speaker's material and economic position. For Del Percio et al. (2017), scholarship on the political economy of language is therefore centrally concerned with how language determines whether and how people can gain access to particular symbolic and/or material resources, whereby 'language emerges as a key site' (p. 55) for determining access to these resources. Analyses in this tradition typically orient more to Bourdieu's conceptualization of capital and, specifically, the linguistic marketplace rather than to Marxist political economy (e.g. Boutet & Heller, 2007; Boutet, 2012; Del Percio et al., 2017). It is this second tradition, orienting more to sociocultural matters, that I align with.

In my book, I do not claim to provide a thorough political-economic analysis of UX writing; what I do offer, however, is a critical discussion of UX writers' status claims in light of their relatively privileged position in the marketplace – both linguistic and more generally. In other words, I am interested in understanding the specific hierarchies that structure contemporary language work (cf. Thurlow, 2020c), and how, say, claiming symbolic capital (for instance in the form of a particular job title) is necessarily also linked to matters of economic capital (such as receiving a higher salary *because of* said job title). In speaking of 'hierarchies of language work' (cf. Thurlow, 2020c), I thus consider both cultural-symbolic as well as material-economic dynamics. It is against this background that I am interested in UX writers as relatively elite wordsmiths; it is precisely their prestigious, behind-the-scenes, and high-impact language work that makes them interesting as an example of the kinds of professionals that shape people's everyday life. My aim with the present book is to document the nature of this work, its (meta)pragmatic organization, and, finally, its cultural-political implications. As such, my book is organized around three key research questions:

1. How is the work of UX writers as elite language workers accomplished and organized?
    - What are the typical professional practices of UX writers; how is this work organized in terms of its regulation and professionalization?
    - What does UX writing, as a high-end form of language work, reveal about the wider hierarchies of language work?
2. How do UX writers themselves understand the role of language in their work?
    - How do UX writers describe their own work and expertise with regards to language?
    - What beliefs about how language works do UX writers make use of in their work, and how do these beliefs impact the interface texts they write?
3. What are some of the cultural-political consequences of UX writers' language work?
    - How do the production circumstances of UX writing influence how interface texts reinscribe normative/ideological discourses?
    - What identities and subject positions do UX writers thus produce and/or privilege through their work?

## 1.4 Methodological Approach

The research that forms the basis of my book was conducted as part of a larger research project titled *Elite Creativities* aimed at investigating the professional practices of an array of elite language workers.[9] In this way, my research design follows that of the *Elite Creativities* project by taking a discourse ethnographic approach (Macgilchrist & van Hout, 2011; see also Blommaert & Rampton, 2011; Krzyżanowski, 2011). As the name suggests, a discourse ethnographic approach integrates ethnography and (critical) discourse analysis as complementary frameworks. This allows for a shift from more textually-oriented analyses towards more situated and actor-related analyses, and aligns with current trends in discourse studies of exploring language/discourse as always situated in specific sociocultural contexts (see Krzyżanowski, 2011). In other words, instead of analyzing only texts-as-products, a discourse ethnographic approach emphasizes the dynamic processes involved in text production, which makes it particularly suitable for my research on the language work of UX writers.

Following Macgilchrist and van Hout (2011), I purposely call my research *ethnographic* instead of calling it *an ethnography*. Ethnography is a contested term, sometimes reserved only for the kinds of sustained, year-long projects typical of conventional, anthropological ethnography. Vis-à-vis such a narrow conceptualization of ethnography, discourse ethnographic approaches align with more recent forms of 'focused ethnography' (Knoblauch, 2005) or 'short-term ethnography' (Pink & Morgan, 2013), which emphasize intensity – of time, data, and analysis – over extensity. It is for this reasons that Macgilchrist and van Hout (2011) describe their own approach as *ethnographic*, using the adjectival form to signal an orientation to the epistemology and research methods of ethnography without over-stating the scope and findings that a discourse ethnographic study can provide. Similarly, my own research is ethnographic, but it does not constitute a conventional ethnography in a narrow, anthropological sense. In fact, given the specific conditions and constraints of the COVID-19 pandemic during which I conducted my research, I inevitably had to find creative solutions for meeting my participants through online encounters, as I will discuss below.

Overall, I understand ethnography not as a fixed method or a set of field-work techniques, but as a '"full" intellectual programme', which entails a particular understanding of humans, society, and ultimately also language (Blommaert & Jie, 2010, p. 5). With its roots in anthropology, ethnography inevitably understands language as 'socially consequential' (Blommaert & Jie,

[9] For details about the *Elite Creativities* project, see www.crispinthurlow.net/elite-creativities/ (accessed 8 October 2022).

2010, p. 5). In other words, from an ethnographic perspective, language is always understood as a resource people use to *do* something in the world. This, of course, is also how sociocultural linguists understand language, who, especially recently, frequently incorporate ethnographically-oriented methods and approaches (cf. Blommaert & Jie, 2010; Krzyżanowski, 2011). Understanding language as sociocultural practice has a number of consequences for ethnographic research, two of which are particularly relevant for the research I present in this book. First, from a (discourse-)ethnographic perspective, communication inevitably entails not only language but also non-linguistic means, since language is always situated in wider patterns of social behaviour. As will become apparent in the subsequent chapters, such a multimodal and often more-than-linguistic understanding of communication is crucial for UX writing. Second, a (discourse-)ethnographic approach tends to centre an emic perspective and focuses on the practices, perspectives, and beliefs of research participants. This, too, was central in my research, where I sought to analyze the work of UX writers primarily in terms of how they themselves understood and made sense of language. I will return to this aspect when discussing my research design; first, however, I account for my own positionality in this discourse ethnographic research.

### *1.4.1 Situating Myself*

The question of access is a key concern in ethnographic research. Not only does access stand at the beginning of most ethnographic projects, but it is also inevitably tied to the researcher's own identity. In my case, my research on UX writing is underpinned and informed by my previous experience of working in the UX industry for almost a decade. Between 2012 and 2021, I worked as a content and communications specialist at a small Swiss UX agency, where I was responsible for anything that had to do with language and communication. My work included only little UX writing, and there was no one else who worked as a UX writer at the company. Still, it was there that I encountered UX writing – as a term and as a professional role – for the first time. Importantly, my experience of working in UX meant that when I started my research, I had the status of an insider, which made it easy (or at least easier) to connect with UX writers for informational interviews. Throughout my research, I also continued to attend and eventually started to organize online community meetings with and for UX writers.

My past experience of working in the UX industry as well as my continuing engagement with the UX writing community certainly influenced my research. For instance, while my analysis of the history and origins of UX writing is based primarily on interviews with practitioners, it is of course *also* informed by my own, lived experience. I distinctly remember the first time I encountered the job title 'UX writer' (in early 2017); it is also through my own experience that I know that the job title 'content strategist' predates it, but the job title 'content designer' does not. (I will discuss the relevance of these various job titles in Chapter 3.) Similarly, my discussion of the ideology of the invisible interface in Chapter 5 emerged from my analysis of interview data, yet throughout this analysis I was distinctly reminded of how, in my own professional introduction to UX, I had been told – and had believed for a long time – that a good interface should be invisible. As such, my past work in the UX industry influenced not only my access to UX writing as a research site, but it also shaped the analyses and the interpretations I present in this book. Having said which, the same would be true if I had had no prior knowledge of UX or UX writing – although differently, this, too, would have shaped my research. There is, as Irvine and Gal (2000) remind us, no 'view from nowhere'; all research always comes from a particular position.[10] In light of this, I find that my involvement in the UX industry has been mostly beneficial to my research, not just in terms of gaining access, but also because it allowed me to more thoroughly focus on UX writers' accounts of their work instead of spending time on explanations of, for instance, UX-specific terms I was already familiar with. As a whole, my conclusions are thus grounded in immersion in the field.

## 1.5 Research Design

In keeping with my focus on practitioners' own, reflexive views, my research was organized into three consecutive stages, each addressing a different aspect of UX writers' language work. As mentioned above, this three-part research design was developed within the *Elite Creativities* project with its common framework. Three core data collection activities were shared: (1) profession mapping, (2) language biographies, and (3) production chain analysis. In short, this entailed moving from the profession in general to language workers' reflexive understanding of language to the realization or application of these understandings in actual production processes. In what follows, I give a brief explanation of each of these data collection stages.

### *1.5.1 Stage 1: Profession Mapping*

I began my research with a method called *profession mapping*, which we took to be a necessary first step in research engaging people's professional practice

[10] While Irvine and Gal make this comment with regards to language ideologies, I believe one can understand their point also as a more general critique of the (false) distinction between 'objective' and 'subjective' research. They are of course not the only ones to point this out; attention to researcher positionality is, for instance, also central to feminist research practices (e.g. Lazar, 2007), which involves much stronger engagement with critical (self-)reflexivity.

(Thurlow, 2022; cf. also Candlin, 2003; Sarangi, 2005). This entails a broad description of the profession itself, establishing, among other things, its history and origins, typical training programs and career trajectories, information about regulatory bodies or otherwise influential institutions, insights into professional associations and/or networks, information about employment conditions and salary scales, as well as a general sense of how people working within this profession organize themselves *as* a profession (or not). In terms of data collection, profession mapping is necessarily a broad and diverse endeavour that differs from profession to profession. In my case, it ranged from archiving industry newsletters to conducting informational interviews with UX writers to attending a three-day training course on UX writing. (I will discuss these different, complementary datasets in more depth in Chapters 2 and 3.)

Profession mapping is useful not least because it enables researchers to (more) clearly define the profession they discuss. Scholars in applied linguistics have long argued that such an engagement with the particular working conditions of practitioners is a necessary condition for research in the workplace. As Candlin (2003; after Cazden, 1970) explains: 'if the researcher cannot enter the world of the practitioner, then the practitioner cannot enter the consciousness of the researcher' (p. 389). In other words, there is a need to first understand the 'world' of a particular profession, especially if one seeks to offer more than a naïve description of professional practices. At the same time, a thorough engagement with the profession sheds light on the structural conditions that impact (language) work. In short, profession mapping is about examining what Bourdieu (1993) might call the (professional) field, including the internal and external power relations that structure and determine the field.

### *1.5.2 Stage 2: Language Biographies*

The second stage in my research entailed the elicitation of what we have called *language biography interviews*. Orienting to work of scholars in second-language acquisition such as Kramsch (2009) and Busch (2006, 2012), these language biography interviews centre practitioners' own perspectives on language as part of their work; this is connected also to their (professional) biographies and career trajectories. In my case, they were conducted as semi-structured interviews and organized around a set of shared questions established within the wider *Elite Creativities* project. To give a sense of the nature of the interviews, some of the questions I asked UX writers were, for instance: *How were you drawn to this type of work? What kind of professional and/or language-related training did you receive? What role do you see language playing in your work?* As such, questions centred interviewees' metalinguistic or reflexive framing of language in relation to their work.

Language biography interviews have traditionally been used for language learning purposes; in our research on wordsmiths, we expanded this notion, using them to elicit another form of 'folk-linguistic' (cf. Preston, 1996) – or simply non-academic – account on language. In this way, the language biography interviews were partly intended to de-centre the academic researcher, bringing to the fore practitioners' own views on how language can and should be used. As Busch (2012) suggests, (language) biographic accounts are particularly suitable for mediating between micro and macro level analyses. In my case, this meant that while a single interview might focus on an individual's experience and perspective, taken as a whole, a corpus of language biography interviews on one profession could offer a unique, practitioner-centred view on the role and understanding of language more generally. This method was thus particularly suitable for tracing how certain beliefs about how language works (or should work) traverse UX writing; related to this, my language biography interviews also served as a starting point for understanding the cultural-political implications of these beliefs.

### *1.5.3 Stage 3: Production (Chain) Analysis*

As a third and last stage in the research design, I initially envisaged conducting a *production chain analysis.* Loosely orienting to the notion of commodity chain analysis, which entails tracing the design, production, and distribution of products (see Thurlow, 2020a), a production chain analysis aims to offer a holistic view on the production of specific texts or semiotic artefacts. Akin to research centring text trajectories (e.g. Woydack & Rampton, 2016; Lillis & Maybin, 2017), this would have entailed following UX writers' work from initial briefing meetings to the eventual publication of their work (or its 'release', as this stage is typically called in UX writing), ideally through fieldwork at a software design company. As mentioned, however, because the start of my research coincided with the outbreak of the COVID-19 pandemic, I had to adapt these original fieldwork plans. While I had envisaged doing on-site fieldwork with UX writers, I had to rely on remote fieldwork – partially because, for a long time, the borders of different countries were closed to outsiders,[11] but also because as a result of local restrictions such as lockdowns or curfews, most UX writers ended up working from home, regardless of where in the world they were located. The conditions of their work had also changed with and because of the pandemic.

In fact, unlike some professions where on-site work was strongly encouraged once governments started lifting pandemic restrictions, many UX writers

[11] To give but one example: the United States, where I initially planned to do fieldwork, imposed a travel ban for almost an entire year (from 31 January 2020 to 7 November 2021).

continued to work partially from home long after restrictions were lifted. In light of this, I adapted my fieldwork towards a form of 'remote ethnography' (Postill, 2016), which entailed observing team meetings of UX writers as well as follow-up interviews focused on UX writers' day-to-day work practices. Orienting to the kinds of reflexive, discourse-based interviews that Macgilchrist and van Hout (2011) consider typical of discourse ethnographic research, the follow-up interviews I conducted involved UX writers discussing a particular semiotic artefact they had recently produced or were currently working on. Such reflexive interviews typically aim to understand writing choices and explore the tacit knowledge and strategies people draw on when writing in professional contexts (cf. Odell & Goswami, 1982); in my case, I used them to also learn more about the broader production processes and different stages that make up the production chain of UX writing.

While I could not conduct a production chain analysis per se – most of my fieldwork was relatively focused and selective due to its online nature – this remote fieldwork nonetheless provided valuable insights into some of the backstage production processes that structure UX writing. At the same time, it helped me revisit some aspects of my initial mapping of the profession, clarifying certain points about the day-to-day work of UX writers that I had started to understand but could not fully explain with the data I had from the profession mapping. Ultimately, then, my book remains underpinned by a decidedly discourse ethnographic approach, one which involved sustained engagement with UX writers, and which allows me to discuss the implications of language in digital media from a production-oriented perspective. To close this introduction, I offer a brief overview of what my production-oriented perspective looked like in practice by outlining the individual chapters of the book.

## 1.6 Overview: The Production, Pragmatics, and Politics of Digital Media

As a whole, my book addresses the production of digital media by centring UX writers' metapragmatic understandings of language and communication. Throughout the course of my substantive chapters, I move from addressing UX writing as a profession to discussing UX writers' reflexive understandings of their (language) work to, finally, a critical examination of the production and products of UX writing. In most chapters, I draw on more than one data set; as indicated above, in my analysis, I triangulated data and insights from usually two or sometimes all three stages of the research. The chapters in the book are thus loosely organized around the three methods/stages outlined above. Beyond this, the chapters are organized into three thematic parts that address the profession of UX writing (Part I); matters of status and hierarchy

(Part II); and, finally, the cultural politics of UX writers' work (Part III). A throughline in the book is UX writers' posthumanist way of working, and how they, without naming it as such, emphasize aspects such as the non-determinacy or the more-than-representative nature of interface texts. To make sense of this, I draw on posthumanist thinking at various points in my analysis; most importantly, though, I draw these threads together in the concluding chapter, where I outline a proposal for a posthumanist approach to analyzing digital media interfaces in order to better address the overall politics and power of such texts.

### *1.6.1 Part I: Mapping the Profession: The Language Work of UX Writers*

In the first part (Chapter 2), I begin by giving a general introduction to the language work and expertise of UX writers. Through this, I show that the work of UX writers is not always easy to pinpoint and sometimes hard to delimit from that of other professionals in software design. I discuss how UX writing is (historically) connected to user experience design and human–computer interaction (HCI), and I begin to outline some of the ways in which this impacts UX writers' work in terms of how they understand language and/or communication. Additionally, I explain in more depth in what way I understand UX writers as elite language workers or wordsmiths (Thurlow, 2020c), who work within the privileged space of the so-called tech industry (cf. Dorschel, 2022). As a whole, this first analytic chapter is somewhat more descriptive than subsequent chapters, but it offers a necessary mapping of the terrain that I build on in the remaining parts of the book.

### *1.6.2 Part II: Establishing Status: UX Writing as Elite Language Work*

In the second part of the book (Chapters 3 and 4), I turn more specifically to UX writers' status as wordsmiths by exploring how, specifically, they claim status and expertise. In Chapter 3, I focus on how UX writers discursively professionalize and legitimize their own (language) work. Drawing on Urciouli's (2008) work on skill discourse and on scholarship in the sociology of professions, I identify 'writing-as-designing' as a unique skill UX writers claim for themselves in order to establish the privileged status of their language work. I argue that the discursive construction of skills can be understood as an 'appeal to professionalism' (Fournier, 1999) that wordsmiths like UX writers deploy in order to claim status and expertise in both local and wider hierarchies of language work.

Staying with the topic of privilege and status, Chapter 4 looks at UX writers' status claims with a particular view to their (dis)avowal of creativity. Orienting to recent work on discursive creativity (Jones, 2010) and rhetorics of creativity

(Thurlow, 2019), I discuss how UX writers strategically deploy claims to *non*-creativity in order to establish their (elite) professional identity. I show that this is connected to an ideological distinction between form and function, whereby UX writers position themselves as concerned primarily with the communicative function of software interfaces and not with their verbal form. A common thread in both Chapter 3 and 4 is UX writers' strong orientation towards UX design, which in turn informs their specific views on how language works – or should work – in digital media interfaces. These deep-seated beliefs in turn influence the kind of texts UX writers produce, as I argue in the third part of the book.

### *1.6.3 Part III: Producing Little Texts: Politics and Power in UX Writing*

In the third part of the book (Chapters 5 and 6), I address how UX writers' understanding of language and communication impact their text production in normative ways. In Chapter 5, I turn specifically to the matter of semiotic and media ideologies (Keane, 2018 and Gershon, 2010, respectively) in UX writers' work. Building on my analysis of non-creativity in Chapter 4, Chapter 5 hones in on invisibility as a presumed ideal in UX writing. Invisibility is a characteristic of much elite language work (cf. Thurlow, 2020c); here, I am specifically interested in the ideology of the invisible interface that permeates UX writers' beliefs about how language should work. I discuss how UX writers' invisible language work has wider, cultural-political consequences as it always privileges a particular position from which things are made (in)visible. In other words, the invisibility UX writers strive for depends on a default or ideal user.

I elaborate on the concept of the ideal user in Chapter 6, where I argue that UX writers' work necessarily also involves stylizing and crafting a certain audience. In this last analytical chapter, I turn, finally, to some of the concrete textual products UX writers create. Examining cookie consent notices as one particular – and particularly prevalent – form or genre of UX writers' work, I discuss how the 'little texts' (Pappert & Roth, 2021) that UX writers produce are both interactionally and ideologically consequential. Specifically, I argue that in their work, UX writers do not only write *for* an audience, but that they also craft an audience through their work, thereby configuring who should and who should not be using digital media. Importantly, the normative aspects of UX writers' work surface not just in the verbal aspects of these interface texts, but also through the dynamic interactions these texts encourage or even demand. It is this that I turn to also in the conclusion of the book, where I reflect on how power and normativity surfaces in UX writing through dynamic and more-than-representational aspects.

### 1.6.4 *Conclusion: Towards a Posthumanist Sociolinguistics of Digital Media*

In the final chapter and general conclusion of my book, I reflect on what an analysis of the production of digital media can reveal about the cultural politics of these technologies. Specifically, I suggest that the work of UX writers reveals how digital media interfaces are normative not just in representational terms, but also through the patterns of usage they create. Revisiting some of my interviews with UX writers, I propose that these 'more-than-representational' (cf. Lorimer, 2005) aspects of digitally mediated communication are something UX writers understand particularly well – better, perhaps, than sociocultural linguists. On this basis, I return to UX writers' own understanding of language and communication, this time not to examine their semiotic ideologies, but to ask what sociocultural linguists might, perhaps, learn from UX writers. Concretely, I propose that the expertise of UX writers offers a starting point for extending the recent turn to posthumanist approaches in sociocultural linguistics (see Pennycook, 2018a). Taking a posthumanist approach to interface texts and attending to more-than-representational, trans-modal, and embodied aspects can in turn enable scholars to deepen their understanding of the overall politics and power of digital media interfaces. In order to understand and – perhaps – eventually challenge the cultural politics of digital media, it is necessary to study not just how people use digital media, but also how digital media are produced, and why they are produced in particular ways. Ultimately, and as I hope to show through the book, looking at the production of digital media and the views of producers is central for understanding how digital media interfaces contribute to the (re)production of status, power, and normativity – sometimes in obvious ways, but often also in more covert and less visible ways.

*I*

# Mapping the Profession: The Language Work of UX Writers

# 2 Designing Words

## The Language Work and Expertise of UX Writers

Today, language is an important part of the work of many – or even most – people. Scholars in discourse studies and critical sociolinguistics have focused on this topic for some time, discussing in particular the role of language work and the language worker in late capitalism (e.g. Duchêne & Heller, 2012; see also Heller, 2003, 2010a). In this chapter, I offer an introduction to the professional practice of one group of language workers: UX writers. As the professionals responsible for creating the linguistic or verbal content of software interfaces, UX writers are a typical example of contemporary language workers – 'the language part of work' (Boutet, 2012, p. 208) is not just part of their work but its main focus. At the same time, UX writers differ from the kinds of language professionals that scholars in critical sociolinguistics have discussed in much of their work, which focuses predominantly on low-status and often exploited workers, or what Heller (2010a) dubs the 'wordforce'. Rather, UX writers are relatively elite language workers or 'wordsmiths' (Thurlow, 2020c), the kinds of professional language workers that are (at least relatively speaking) privileged and whose work is valued more highly, often in both symbolic and material terms. In what follows, I offer an introduction to the work of UX writers by discussing the linguistic (and more-than-linguistic) practices that constitute their language work, their status as wordsmiths, as well as the way they understand and theorize language as part of their work.

In simple terms, UX writers are responsible for writing user interface texts, that is, the words people see when they use digital media (though, as I will discuss later, the actual work UX writers do is both more comprehensive and less easily demarcated.) While scholars in digital discourse studies have frequently taken up interface texts as objects for critique, there is usually only little consideration given to the linguistic work that goes into producing these interface texts. In scholarly discussions of interfaces and interface texts, UX writers are frequently treated as an unknown entity, and claims about their role are made based on interpretations of the outcomes of designers' work, that is, the products rather than the producers. By the same token, critical sociolinguists often tend to discuss language workers as (primarily) objects of study while disregarding their linguistic expertise as a valid – or at least potential – source of sociolinguistic theory. In this regard, Thurlow and Britain (2020) point out that as academic experts on language, sociocultural linguists often have 'an underlying sense that other workers ought somehow to be orienting to our expertise' rather than asking about *their* expertise and how they might impact or extend sociolinguistic understandings of language (pp. 74–75; see also Thurlow, 2018). It is with this framing in mind that I approach the profession of UX writing in the current chapter. Following the example of Thurlow and Britain, I am interested not just in documenting the professional practices of UX writers but also in understanding how they themselves theorize language and communication as part of their work.

The current chapter is structured as follows. I first discuss the – admittedly sparse – engagements of academics with UX writing. This survey of past work makes one thing clear: while scholars have discussed at large how *users* interact with digital media, there is considerably less work on the *producers* of digital media. After this brief survey, I offer an in-depth introduction to UX writing in three parts. First, I offer a broad mapping of the profession, where I discuss the (brief) history and origins of UX writing. Next, I give a general introduction to UX writers' language work, illustrating how their work is both centrally concerned with writing interface texts and much more complex than that. Finally, I turn to some initial ethnographic observations about UX writers' concrete text production. In this last regard, I am particularly interested in how UX writers mobilize (meta)linguistic knowledge in their work, arguing that they are not just language workers but also language experts. Overall, this chapter thus offers a first description of the work and profession of UX writers, orienting primarily to these wordsmiths' own views and understandings of what it is that they do. I start, however, with a brief survey of academic perspectives on UX writing.

## 2.1 Academic Perspectives on UX Writing

Given that the job title 'UX writer' has gained popularity only in the last decade, it is hardly surprising that there is not much sociolinguistic research – or any research for that matter – on UX writing. Thus, while substantial research has been conducted on how people communicate with and through digital media (see, e.g., the edited volumes by Thurlow & Mroczek, 2011; Herring et al., 2013; Jones et al., 2015; Georgakopoulou & Spilioti, 2016), few scholars consider the role of software design in shaping how people (can) communicate, and even fewer discuss the role of the professionals creating these interfaces. Notable exceptions include the kind of work spearheaded by Djonov and van Leeuwen (2013, 2017), whose work on PowerPoint discusses the normativity of software and how software as semiotic technology significantly shapes what people can do with and through software. While their work does not explicitly discuss interface texts, it does, inevitably, touch on linguistic elements in the interface design of PowerPoint, such as the impact of specific labels and how, for instance, the label 'Title Slide' implies that users should, in fact, create a slide with a title. Similarly, Kvåle (2018) discusses the semiotic conventions of Microsoft Word templates, arguing that pre-defined verbal resources in such templates impact users' identity expression. Again, Kvåle does not explicitly focus on UX writing, but her analysis nonetheless touches on how interface texts guide users' choices. However, despite these advances, research in this tradition still treats interfaces as objects of analysis rather than sites of language work; as such, interface texts are studied as linguistic or semiotic products rather than in terms of their production.

There are a few contributions in a recent edited volume by Jones et al. (2015) that address the role of software producers at least implicitly by discussing the construction of people's online selves as 'a jointly negotiated, interactive process between designer and user' (Hafner, 2015, p. 110; see also Chik, 2015 and Merchant, 2015 in the same volume). Yet even here, designers are treated as an anonymous entity, and claims about their role in this 'jointly negotiated process' are made based on finished designs rather than on observed actions or statements of designers. Additionally, the complex constellation of different people and professions involved in software design is typically distilled into the simple, abstract notion of 'the designer'. Consequently, the main focus lies on the users rather than the producers of these interfaces.

A rare exception is the work of media and communication scholar Massanari (2007, 2010), who explicitly centres producers of digital media in her work. Specifically, she examines the discursive development of the professional practice of information architecture. Despite its rather different name, information architecture is a profession that is quite similar to UX writing. While UX writers write the linguistic content of software, information architects are usually responsible for structuring and classifying that content, for instance by defining navigation schemes, taxonomies, or controlled vocabularies.[1] As I will discuss in this chapter, this work certainly overlaps with UX writing; indeed, some UX writers, including some of my interviewees, would claim that information architecture is part of UX writing (or vice

[1] The term 'information architect' has somewhat fallen out of fashion in the last decade and is now often replaced by or conflated with the term 'UX designer'. In fact, UX design is often seen as an umbrella term that encompasses all kinds of professions and practices related to the visual but also conceptual design of software (more on this below). Based on my personal experience, I would suggest that the term 'information architect' is also more US-centric and used less often in Europe.

versa). Massanari's research bears many similarities to my own project – she, too, has previous experience working in the professional field she discusses, allowing for a very distinct view on the practices information architects engage in. However, as she acknowledges, her work falls short of directly observing the production processes of software design, though she highlights the importance of exactly this kind of work:

> While we have often focused on the social and cultural impacts of technological artifacts, we have remained relatively silent as to the ways in which professional practices and organizational behaviors shape these products. Tracing the development of an artefact . . . would be beneficial in understanding the very real consequences these practices have on the design and use of new technologies. (Massanari, 2010, p. 412)

As Massanari points out, while scholars have written and talked much about the use and impact of technological artefacts, they have still paid only little attention to how these artefacts and texts come to be. Like her, I suggest that focusing for once not just on the use but more strongly on the production of interface texts can provide a valuable complementary perspective not just on digital media but also on the conditions of UX writing as a new, elite form of language work. With this in mind, I now turn to a cursory overview of the language work and practice of UX writing.

## 2.2 The Data

For the initial overview of UX writing that I provide in this chapter, I draw on a variety of complementary sources. First and foremost, I base my discussion on interviews with UX writers. Over the course of my research, I conducted 21 semi-structured, informational interviews with people who work in UX writing. Each interview was conducted between November 2020 and April 2021 via video call and lasted between 45 and 60 minutes, with questions addressing interviewees' educational and professional trajectory, their everyday work, their views on the profession as a whole, and, specifically, their use and understanding of language in their work. All of my interviewees worked at medium to large international companies, and most were based in the United States, with more than two-thirds of them based specifically in California. This choice of site is no coincidence – the United States, and, in particular, Silicon Valley, is a hotspot for trends in software design. Likewise, while smaller companies have started hiring UX writers in recent years, it is major international software companies such as Google or Dropbox that have popularized the job title 'UX writer' and contributed to its appeal and prestige in the industry.

Second, my research is informed by a close reading of key professional texts on UX writing. In this regard, I have examined three books written by practitioners and directed at other (potential) UX writers:

- *Writing Is Designing* by Michael J. Metts and Andy Welfle (2020)
- *Strategic Writing for UX* by Torrey Podmajersky (2019)
- *Writing for Designers* by Scott Kubie (2018)

From my engagements with UX writers, I know that these three books are well-known in the UX writing community and frequently recommended to newcomers as a starting point; they were also mentioned directly by several of my interviewees. Beyond this, I also draw on a variety of less official and at times more ephemeral secondary material produced by UX writers about their own work. This includes podcast episodes, newsletters, blog articles, and social media posts of UX writers discussing their day-to-day work, their career trajectories, as well as more general aspects of the profession, such as suggestions for best practices.

Third, and in order to discuss UX writers' concrete, day-to-day work, I draw on fieldwork I conducted at Novus,[2] a large international software company, in spring 2022. I conducted this research under a non-disclosure agreement; for confidentiality reasons, I thus cannot offer a more concrete description of the company itself. I interacted with one UX writing team at the company, which consisted of nine people. Due to the ongoing COVID-19 pandemic, all of this fieldwork was conducted via video call (the team worked primarily from home at the time). Over the course of two months, I was able to observe and record two different types of meetings:

- Two team meetings of the UX writing team, which function as a bi-weekly feedback session where all twelve UX writers discussed results or problems related to their individual work
- Four meetings between individual UX writers and two to four designers/project managers, where the participants discussed the concrete, ongoing work they conducted together

After these observation sessions, I conducted follow-up interviews with six different UX writers. I asked the interviewees to bring one or two examples of their current work to our video call in order to exemplify some of the work I had heard about. The data I discuss in this chapter stems primarily from these follow-up interviews; due to the non-disclosure agreement, I focus my discussion on general statements rather than specific examples of these UX writers' work.

My investigation of UX writing as a whole is also underpinned by my own personal experience of having worked in the UX industry for nine years (between 2012 and 2021), as well as my continuing engagement with the UX writing community, which includes attending as well as organizing online

[2] The name Novus is a pseudonym.

community events with and for UX writers. It goes without saying that my own involvement in the UX writing community is what piqued my interest in studying the language work of UX writers; it also served as a valuable entry point into local and global UX writing communities, which in turn allowed me to gain access to some of the most well-known professionals in the industry for informational interviews. In all my interviews with UX writers, I disclosed my background to the interviewees but made sure to (truthfully) position myself as someone who knows the profession without being an expert (my past work experience included only little UX writing work). This way, my interviewees remained positioned as the experts on the topic. At the same time, presenting myself as someone with (some) experience in the field of UX guaranteed that my interviewees could rely on common terminology and a basic understanding of, for example, other professional roles UX writers typically collaborate with. My past experience in the UX industry certainly influenced my analysis. However, I believe that the benefits of this outweigh the downsides, as it allowed me to more thoroughly engage with the topic, especially with regards to the overall emergence and development of the profession of UX writing – which is what I turn to discussing next.

## 2.3 Mapping the Profession

### *2.3.1 UX Writing as an Emergent Profession*

'Look at your favourite app, site, or interface. Take all the words away and what do you have? Not much!' (Kubie, 2018, p. 2). It seems obvious: all software interfaces contain words – indeed, words make up a large part of them, as Scott Kubie, a well-known figure in UX writing, points out on the first pages of *Writing for Designers*. And yet, a decade ago, it would have been hard, if not impossible, to find a person with the job title 'UX writer'. This is not to say that there were no words in software interfaces, but it would have been rare to find someone whose job revolved solely and explicitly around writing the small pieces of text in software interfaces. While it is impossible to pin down an exact year in which the job title 'UX writer' emerged, I became aware of it in early 2017 when talking to a UX design consultant from Belgium, who excitedly shared a recent job opening for a UX writer at Amazon. (Clearly, the prospect of UX writing as a profession was new to her as well.) My interviewees similarly tended to place the emergence of this particular job title in the mid-2010s. As a case in point, I offer Extract 2.1, where one of my interviewees, Scarlett, shares how she got into UX writing.[3]

[3] Throughout this book, I use pseudonyms for all my interviewees. Further identifying information is either anonymized (e.g. [international software company]) or entirely omitted.

Scarlett is an established in so far as one can use the term within UX writing – UX writer working in the United States.

**Extract 2.1: Informational interview with Scarlett**

I talked to a manager at [international software company], and this was in 2010, he was the writing manager, he managed the team of technical writers who wrote developer documentation. . . . He also managed a very small team that worked on the consumer-facing content, and he said, "I can't put out a job description for this job. I need somebody writing the words on the screen that the nine-year-old will see on Christmas morning and understand how to set up [product the company developed] all by themselves. . . . How do you write that job description?" UX writer was not something that existed a decade ago. So, he s– but he said, "*You* can explain complex abstract concepts . . . you're gonna be great at this."

In Extract 2.1, Scarlett explains how the job title 'UX writer' was not only something that did not exist in 2010 but that it was something her future manager could not even put into words in order to publish a job opening (lines 4–8): 'How do you write that job description?' (lines 7–8). Similar to how writing in other professions is often not consciously recognized as such (e.g. Opel & Hart-Davidson, 2019 on writing in care clinics), here, too, writing is (or was) misrecognized. While there were some people doing similar work – though not many, as Scarlett indicates (only 'a very small team', line 3) – conceptualizing this kind of language work as its own profession was, it seems, impossible. By contrast, at the time of writing, there are over 250 job openings listed on the job portal *indeed.com* for the search term 'UX writer'. And while the boundaries of the profession are somewhat vague – what some call a UX writer, others call a content designer, and what some consider to be part of a UX writer's daily work, others see as the responsibility of an information architect – there is general agreement that UX writers are, broadly speaking, concerned with the linguistic part of software interfaces. As a starting point for identifying what this work entails, I turn to Extract 2.2: this is how UX writing is defined by Google on the *Google Design* career site.

**Extract 2.2: Google's definition of UX writing[4]**

UX Writers advocate for Google design and help shape product experiences by crafting copy that helps users complete the task at hand. They set the tone for content and drive cohesive product narratives across multiple platforms and touchpoints. As our resident wordsmiths, they work with a variety of UX design related jobs including researchers, product managers, engineers, marketing, and customer operations to help establish connective language and a unified voice.

[4] Source: *Google Design – UX Writer*. Retrieved from https://design.google/jobs/ux-writer (accessed 28 October 2022). As snapshots by the Internet Archive's *Wayback Machine* confirm, this text has not changed since the website's publication in 2017.

Apparently, being a UX writer at Google quite literally means taking on the identity of a 'wordsmith' (line 4). And indeed, from this glossy description, one gets the impression that this profession is not simply about churning out words but about doing more important and more exciting things with these words. UX writers, readers are told, 'craft[] copy' (line 2), work on 'connective language' (line 6), and establish 'narratives' (line 3) through their writing. Through this, the description of UX writing in Extract 2.2 seems to evoke artistry and craftsmanship, attention to detail, and a skilful, perhaps even creative, approach to 'the language part of work' (Boutet, 2012). Evidently, a job description listed on a career website is a quite particular, promotional text genre, meant to attract future employees. Nonetheless, this text captures the kind of cachet that comes with the job title 'UX writer', which certainly bears none of the stigma associated with other kinds of language work, such as, for instance, the work of call centre agents (cf. Woydack, 2017). Indeed, the UX writers I encountered through my research were all quite proud of their work and expressed strong personal identification with what they were doing. Overall, UX writing is a profession that is highly valued; this is a job title one can be proud to claim.

At the same time, UX writing as a profession is fairly well remunerated. An industry survey published by the *UX Content Collective* in 2019 reports, for instance, that the average salary of a UX writer in the United States is USD 126,000. Other industry surveys report similar numbers.[5] It is likely that these high numbers are in part due to the high salaries paid in Silicon Valley (38% of the US-based respondents in the *UX Content Collective*'s survey reported being located in California); there are, quite likely, also simply more jobs for UX writers in that area. In any case, and as the *UX Content Collective* concludes, '[i]t's clear that UX writers and content strategists are well-paid. Even outside of major tech markets, salaries are well above market average in the United States (USD $56,516, according to the 2015 census)'. From my interviews and my engagement with people in the UX writing community, I have also learnt that UX writing tends to be better paid than other kinds of language work in the same industry, such as copywriting positions at software companies.

In terms of symbolic prestige, UX writers also at times invoke a moral hierarchy vis-à-vis other language professions, where UX writing is seen – or positioned – as morally superior. Some of my interviewees suggested that

[5] Sources: *UX writer salary survey: How much money do UX writers make?* Retrieved from https://uxwriterscollective.com/how-much-do-ux-writers-make-salary-survey-revealed (accessed 21 July 2022); *UX Writing Worldwide Report.* Retrieved from https://dxrc.qrplanet.com/l/media?f=2ad36b665a91e877fe5e40b35d19a102 (accessed 21 July 2022); *UX Writing Salary Survey 2021 Results.* Retrieved from https://uxwritinghub.com/ux-writing-salary-survey-2021-results/ (accessed 21 July 2022).

while UX writing is about 'helping users', copywriting and marketing are simply about selling things. Of course, in practice, things are never quite as simple. The boundaries between UX writing and adjacent professions are often less well defined, and regardless of the glossy definition of UX writing that Google gives (Extract 2.2), UX writers themselves often describe the nature of their work in more modest terms, highlighting and valuing other qualities, such as analytical thinking, precision, and clear writing. To understand this tension, it is useful to disentangle the arguably short history of UX writing by examining the attribute 'UX' more closely.

### *2.3.2 The Language Part of UX Design*

UX writing is generally seen as a professional specialization within the broader field called 'user experience' or 'user experience design'. Job openings for UX writing positions therefore frequently list knowledge of UX processes as a required skill. In a similar vein, the job profile on the *Google Design* career page (Extract 2.2) continues with a short, two-question interview with a UX writer, whose advice to potential candidates is to '[c]ome prepared with expertise beyond writing. Learn the basics of design, information architecture, or frontend development'. Design in this case refers to the (visual) design of software interfaces, while information architecture refers to how the content of the software is organized. Frontend development, finally, refers to the technical programming of the interface, that is, how the visual design and structural logic are translated into software code. As a whole, this statement points to the highly collaborative nature of UX writing; moreover, it highlights the profession's close connection to the diffuse practice of user experience design.

User experience (UX) is commonly associated with web and software design, but both as a concept and as a practice, user experience is diverse and multifaceted (cf. Bolchini, 2015). Definitions of the term range from 'all aspects of the end user's interaction with the company, its services, and its products' (Norman & Nielsen, 2021, para. 1) to 'a momentary, primarily evaluative feeling (good-bad) [that someone has] while interacting with a product or service' (Hassenzahl, 2008, p. 12). In fact, as Turner (2017, p. 2) points out, many researchers have simply avoided defining the term. Though most definitions do not limit UX to digital technology, it is typically associated with digital media. In practice (and this is something I can confirm based on my own experience), the term UX is often used as what Bolchini (2015) calls 'a more colorful definition of "web design"' (p. 234), and it is used as a buzzword, especially by business professionals. These varying definitions of UX can in part be attributed to the fact that the term has roots in several different disciplines, which in turn emphasize different aspects that have come to be subsumed under 'user experience' as an umbrella term. According to

Jensen (2013), UX has precedents in HCI (where the term is often used with reference or in contrast to 'usability'), in marketing (where it is frequently subsumed under the broader label 'customer experience'), as well as in design (where its roots lie in practices associated with 'user-centred design'). To complicate things further, the term is used by academics (for instance, in HCI or design studies) as well as by practitioners, who might emphasize different aspects or traditions.

Since I am interested in UX writers as practitioners and in understanding their own, reflexive views, I align my understanding of UX with the conceptualizations that are common in the web and software design industry. In this regard, the professional practice most strongly associated with UX is design. One often sees the term 'UX design' (rather than only 'UX'); by the same token, 'design' is often understood as meaning 'UX design' rather than, say, industrial design. That said, within the industry, UX and UX design are generally seen as umbrella terms encompassing a variety of professions. For instance, Google's overview of different design careers (cf. Extract 2.2) lists not only 'UX writer' as a career path but also 'UX engineer', 'UX researcher', and 'UX program manager'. (The list does not entail 'UX designer', though it does feature more specific design-oriented roles, such as 'visual designer' and 'interaction designer'.) From this perspective, and as a working definition, one might say that UX professionals are people who work in web and software design and who are concerned in particular with shaping how users will experience their interaction(s) with digital media; this professional field can be (and frequently is) called 'user experience design'. UX writing, then, is a specialization within this professional field, with UX writers focusing specifically on the language part of UX design.

In fact, many UX writers draw on the connection between UX and design when explaining what UX writing is. For instance, in her introductory book *Strategic Writing for UX*, Torrey Podmajersky (2019) defines UX writing as 'the process of creating the words in user experiences (UX)' (p. ix). Similarly, Michael J. Metts and Andy Welfle (2020) explicitly title their book *Writing Is Designing,* telling their readers to '[s]top writing clever copy. Start writing to design' (p. xi), and Scott Kubie (2018) emphasizes on the first page of *Writing for Designers* that "[w]riting is part of design" (p. 1). All three of these texts stress the broader professional context and practice of UX writing: it's not about writing words in interfaces but about writing (or rather designing) 'words in user experiences' (Podmajersky, 2019, p. ix).

Throughout their introductions to UX writing, the three books also highlight the relative newness of the profession and that words in software design have often been overlooked in the past. While words have always been part of software interfaces, it is certainly interesting that the task of writing these words has only now received attention as 'proper' work within the web and software design industry. When I started my research, a colleague of mine, who had been working in software design for roughly two decades, offered the interesting observation that the rise of the industry's emphasis on language and writing seems to coincide with the rise of a design trend called 'flat design'. While early web and software design was characterized by skeuomorphism, that is, design whereby the GUI is meant to mimic a real-world counterpart through the impression of three-dimensionality and texture, the 2010s gave rise to so-called flat design, which, rather than emulating three-dimensional shapes, is characterized by simple, solid shapes that give an appearance of flatness. (Microsoft's design for Windows 8 is a good example of this and often said to mark the start of this design trend; see, e.g., Page, 2014; Bollini, 2017.) In social semiotic terms, one could say that skeuomorphic design has high naturalistic modality, whereas flat design has low naturalistic and high abstract modality (see Kress & van Leeuwen, 1996/2006). Flat design thus represents a turn in digital design leading from design that imitates physical objects to more abstract design. Perhaps the reduced emphasis on visual elements made room for the words in software interfaces to be foregrounded – or perhaps flat design made it necessary to add more explicit textual explanations to software interfaces. Whether accurate or not, the connection between these two developments is an interesting observation to make and one that shows that UX professionals like my colleague are very aware of the multimodal nature of the digital texts they create and how, for example, visual design and verbal text are always interconnected – a point that my interviewees also frequently highlighted.

Looking to research in critical sociolinguistics, I see another explanation for the increased interest in the language part of software interfaces, which is connected to the rise of websites and apps as commodities in the contemporary consumer market. As Heller (2010b) points out, the increasing importance of and demand for language work in late capitalism has surfaced in part due to an (over-)saturation of markets. Heller observes that in this regard, language can become a source of additional distinction that may set one mass-produced product apart from another. The importance of such symbolic aspects of language is something that Kelly-Holmes (2020) observes quite pointedly with regards to the use of multilingualism in marketing. Speaking of the 'linguistic fetish' (Kelly-Holmes, 2014), she shows that in the linguistic business of marketing, languages are often used not for instrumental purposes but 'to enhance the semiotic landscape of the advertisement' (Kelly-Holmes, 2020, p. 38). Through its symbolic meaning, language can thus become a source of added value in the saturated markets of late capitalism. In fact, the idea of language as added value is a rhetoric that UX writers themselves (or the companies they work for) also employ when trying to convince others of the value of their work. For instance, one of the first prominent public talks about

UX writing, given in 2017 at Google I/O (a popular annual developer conference organized by Google in Mountain View, California), is tellingly titled *How Words Can Make Your Product Stand Out.*[6] It seems that language, here, too, is a source of 'symbolic added value' (Heller, 2010b, p. 103). This added value, in turn, has given rise to the profession of UX writing: today, UX writing is recognized as a distinct role within the software design industry. That being said, its boundaries as well as how one can become a UX writer are matters that are often less clear-cut.

### 2.3.3 *(A)Typical Career Trajectories*

Given the relatively recent emergence of UX writing as a profession, it is not surprising that there is no single, fixed path to becoming a UX writer. More often than not, my interviewees prefaced their career stories with a caveat that their own career path was probably rather untypical. Several of them joked that they became a UX writer by accident. Some became UX writers only because a colleague or manager had pushed them in the right direction (recall Scarlett's story from Extract 2.1). A more junior UX writer I interviewed explained to me that before her current position as a UX writer (which she took up in 2019), she hadn't even known the profession existed. Indeed, she admitted that if the job posting had been for a 'UX writer' rather than a 'junior writer', she probably wouldn't have considered applying. Another interviewee quite adamantly declared that there was simply no such thing as a UX writing background.

While there is no common or required training for becoming a UX writer, what most UX writers seem to have in common is an educational background in a language-related field. Most of the people I interviewed had an undergraduate or graduate degree in English, journalism, or media and communication studies. A cursory overview of job postings for UX writers shows that today, most companies require applicants to have a degree (usually an undergraduate/bachelor's degree) in one of these fields. In terms of training, there are also several online courses and training programmes specifically dedicated to UX writing that have emerged from 2018 onwards, such as a ten-week *UX Writing Fundamentals Course* offered by the *UX Content Collective,* a six-month *UX Writing Academy* offered by the *UX Writing Hub*, as well as a host of shorter workshops and courses. However, completing such a course does not seem to be necessary to get a job as a UX writer. As a case in point, only two of my interviewees mentioned having completed one, and only one of them saw it as an entry point into UX writing. Indeed, most of my interviewees received their training 'on the job', and several of them found their first UX

[6] Source: *How Words Can Make Your Product Stand Out (Google I/O '17).* Retrieved from www.youtube.com/watch?v=DIGfwUt53nI (accessed 16 July 2022).

writing position by transitioning to UX writing from another position at a software company.

When it comes to applying for a UX writing position, being able to list a well-known company such as Google or Meta as a previous employer on one's resume could be decisive, as one of my interviewees explained. By the same token, this may make it hard for newcomers to find their first job in UX writing. One of the least experienced UX writers I interviewed told me that she had been interviewing for over a year until she could sustain herself only through UX writing; which in this case did not mean having a fixed, full-time position but rather several individual freelance contracts that added up to 40 hours of paid UX writing (i.e. the equivalent of a full-time position). Thus, while there are no hard criteria for becoming a UX writer, not everyone can easily get a foothold in the professional field – maybe also because there is no clear career path for this profession.

More broadly, the diverse range of career trajectories my interviewees shared also indicates that it is, in fact, quite difficult to define or delimitate what UX writing is, perhaps because of its relative newness, but perhaps also because, as Thurlow and Britain (2020) suspect, 'this kind of ambiguity and fuzziness of boundaries is actually common for many domains of wordsmithing' (pp. 69–70) – it certainly is the case for UX writing. This is in part also because of the particular, at times ambiguous, and often distributed nature of the work UX writers do, as I will discuss next.

## 2.4 The Work and Craft of UX Writers

### *2.4.1 (More Than) Writing Little Texts*

The perhaps most emblematic kind of linguistic product UX writers create as part of their work is so-called microcopy, sometimes also referred to as 'interface copy' or 'UI copy' (short for 'user interface copy'). Figure 2.1 offers an example of this, showing a screenshot of *Amazon's* check-out process. All the little bits of text one can see in this interface would be considered microcopy; from the navigation at the top down to the two options users can select: the discreet 'Continue and don't gain Amazon Prime benefits' and the more prominent yellow button stating 'Use now – pay after'. Yet, while some UX professionals define UX writing as pertaining only to the writing of interface texts like the ones in Figure 2.1, most UX writers see their work as broader, arguing that UX writers also spend a significant amount of their time planning the overall strategy of these small texts. This may include high-level tasks such as defining (and documenting) the 'voice' of a specific app or website – that is, the imagined personality or 'persona' the company wants users to attribute to the software and that UX writers will seek to express

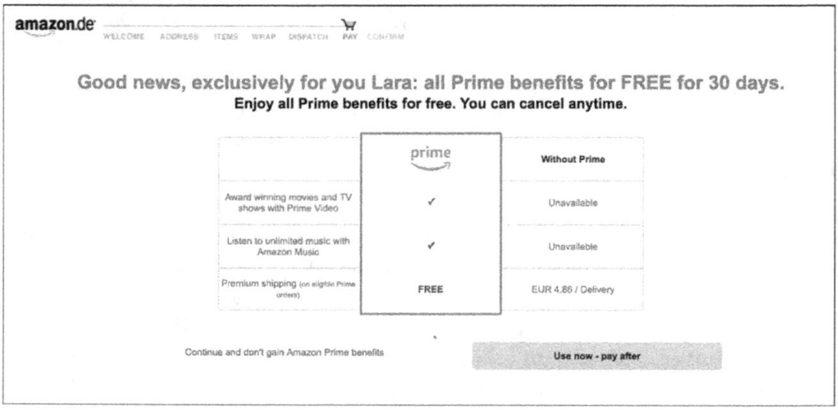


**Figure 2.1** UX writing on Amazon

through their writing (cf. Droz-dit-Busset, forthcoming). It may also include more concrete tasks such as defining the narrative flow of a piece of software and outlining the different stages users will move through when using the software. There are numerous techniques and frameworks UX writers use to document this 'overall strategy of the content', as one of my interviewees called it. To name a few, this might entail:

- *Tone maps*, which 'map' the intended tone of a software's interface text across a specific process users will click through, such as checking out on Amazon
- *User story maps*, which similarly document a process with a particular emphasis on the imagined experience of a prototypical user
- *Style guides*, which document the register or style to be used and typically include not just instructions on spelling but also guidance on how to write in the company's or product's 'voice'

A UX writer's work often also includes doing research, such as background research to contextualize microcopy (for instance by talking to subject matter experts and/or to users themselves) as well as research aimed at evaluating the success of microcopy, for instance through so-called usability testing of interfaces. In usability tests, UX writers – often together with dedicated UX researchers – evaluate a preliminary version of the software with users to find out whether users understand how to use the software and to identify specific issues that should be changed before the software is released. In this regard, the entire second chapter of Metts and Welfle's (2020) introductory book is dedicated to the topic 'strategy and research'. As they explain, 'finding the right words also requires researching your users and testing your ideas with them' (p. 28). Similarly, Podmajersky's (2019) book includes a dedicated section on 'researching UX content' (pp. 110–113) situated in a chapter on measuring the effectiveness of one's writing. The methods that both books discuss are similar and include, for example, interviewing and contextual inquiry (a form of participant observation), but also more specific methods such as usability testing, co-design exercises, as well as A/B-testing, whereby two different versions of a website or app are released to two groups of people in order to see which of the two versions 'performs' better (i.e. is more likely to lead to the desired outcome, which might be for users to sign up for a service or click a specific button).

Especially more quantitative research methods such as A/B testing highlight that UX writers are often required to justify their writing choices by backing them up with 'hard' facts and data. As one of the UX writers I interviewed explained, 'you have to have evidence for everything, you have to do research studies just to prove that you used the right verb here or that people are going to understand exactly which term you used in this context'. While this statement might be a slight exaggeration, it nonetheless illustrates the kind of pressure that is put on UX writers to prove their worth. As the same interviewee pointed out, 'there is this constant struggle of distinguishing yourself as, like, uniquely better at writing than other people ... but also capable of doing – functioning more broadly'. Even if they wanted to, UX writers could perhaps simply not afford to do 'just' the writing.

Two things become apparent from this cursory overview of the work UX writers do. First of all, UX writing entails a variety of processes, few of which would typically be considered writing. Indeed, as one of my interviewees observed, 'the actual writing is like 10% of what I do', suggesting that she spent much more time coordinating, planning, and documenting than writing individual words. Thus, the microcopy UX writers create is perhaps best seen as simply the final outcome rather than the core of their work. Second, and related to this, these processes that UX writers engage in are typically dispersed over a variety of places and people. In this sense, the language work of UX writers is also highly distributed.

### 2.4.2 *Distributed Work*

While there is little – if any – academic work on the language work of UX writers, the work of technical writers is relatively well researched. Technical writing is similar to UX writing, with some people seeing it as a precursor to UX writing, though the two are usually treated as two distinct professions. Unlike UX writers, who write the microcopy users will read, technical writers tend to write texts that are more technical in nature and directed at software engineers as primary addressees, such as software documentation. Still, research on technical writing gives an initial indication of the working conditions of UX writers. Here, scholars of technical communication observe that technical writing is highly distributed: it is characterized by re-use, re-cycling, and the combination of existing texts more so than the writing of new ones (e.g. Slattery, 2007; Swarts, 2010), with writing processes 'tak[ing] place across temporal, spatial, personal and system boundaries' (Hart-Davidson et al., 2006, p. 70). Given the close relationship between technical writing and UX writing, it is not surprising that the work and writing of UX writers, too, is characterized by such distribution and fragmentation.

The distributed nature of UX writers' language work is partially due to the fact that UX writers rarely work on their own. UX writing, like most work in software design, is highly collaborative, with UX writers constantly working with designers, software engineers, researchers, project managers, and others. As a result, their language work is closely affected by visual aesthetics, technical constraints, legal factors, and business goals. Not least owing to this, UX writers tend to understand their work as highly multimodal, and as concerned not just with language, but also with everything that is going on besides language. In practical terms, however, the collaborative nature of their work also means that UX writers are often not fully in control of the words that make it into released software. They may only work on some of the words (this is almost always the case for larger software projects), and they may only work on these words at one stage in the project, for instance, by editing that which someone else has written earlier. Indeed, one of the main pain points my interviewees mentioned was not being involved enough (or early enough) in team meetings with their non-UX writing colleagues. Owing to this, their impact on interfaces may be more limited than that of their colleagues.

UX writers also often work on more than one project at the same time – certainly more so than their colleagues in design and engineering. In this sense, their work entails switching between projects and texts in parallel and distributing their work over time. Finally, their work is also distributed in terms of how the texts they produce are created, stored, and eventually published. For instance, a UX writer may start drafting a narrative flow using Post-It notes, then move to writing microcopy in a design file, before the text is eventually stored in the software's code. Throughout this, the text may be re-shaped by visual or technical factors: changes in the visual design may force UX writers to change the length of their text, or they may be required to formulate an error message more vaguely so that it can be used in multiple contexts. What emerges across all of this, then, is that the language work of UX writers is typically distributed or dispersed across space, time, and multiple people. The agentive force, then, is also not situated in an individual person or agent, rather, it is distributed across this network or assemblage that comprises a multiplicity of people and material-discursive artefacts.

It is exactly due to its distributed nature that UX writing cannot be explained by looking only at the linguistic products these wordsmiths produce. In order to understand the nature of this work as well as the role of language within it, it is necessary to understand the production processes going on behind the scenes, including all the work that happens before – or besides – the final text. Such more discourse ethnographic investigations of how people work with and through language can be invaluable for understanding the linguistic products language workers produce and for tracing how production circumstances impact and shape language workers' linguistic output (see, e.g., Macgilchrist, 2020 on language work in the publishing industry or Jacobs, 2020 on public relations professionals). With this in mind, I now turn to a preliminary discussion of UX writers' concrete text production, focusing specifically on UX writers' (meta)linguistic knowledge and on how they conceive of language in their day-to-day work.

## 2.5 The Linguistic Expertise of UX Writers

### *2.5.1 Showcasing Linguistic Expertise*

To examine the day-to-day work of UX writers, I draw on my fieldwork at Novus. As a reminder, all of this fieldwork was conducted via video calls; it involved observations of online team meetings with UX writers, designers, and project managers, as well as follow-up interviews during which UX writers talked through the linguistic products or outputs they had produced in past and ongoing work. In my discussion here, I draw primarily on the fieldwork I conducted with one single UX writer, Lev, with whom I had the most prolonged contact. I attended several meetings that Lev had with designers, observed him discuss part of his work at a team meeting with UX writers, and I conducted a one-hour follow-up interview with him.

For the follow-up interviews, I had asked the UX writers from Novus to bring one or two artefacts of their work to the interview. While some of them shared and discussed their work based on visual design files, another popular semiotic artefact that interviewees brought to the meeting – and the kind of document I discussed with Lev – was a so-called copy doc: a simple rich text document that contained all the microcopy pertaining to a particular project or task, along with extensive comments about these texts. These comments included things such as alternative or previous versions of the interface text, notes about its context (and/or screenshots to visualize its immediate placement in the interface), open questions (from the UX writer and/or their colleagues), as well as the UX writer's explanation for their specific revisions or recommendations. Typically, this would be realized either as a text document that could be shared via co-editing services such as Google Docs or

Dropbox paper or as a wiki page.[7] As a way of illustrating this, Figure 2.2 offers a schematic representation of a copy doc; note, however, that copy docs would often be much longer than this and consist of multiple pages.

One thing that these copy docs illustrate nicely is that the 'little texts' (Pappert & Roth, 2021) UX writers produce are much bigger than their mere visual size would suggest. While the final microcopy might be only a few words long, UX writers certainly produce a lot of additional text about these 'output' words, and it is this extensive metalinguistic commentary that interests me. Extract 2.3 offers an example of the kinds of explanatory notes that might be included in such a copy doc (this corresponds to column 5 of the schematic in Figure 2.2). Concretely, Extract 2.3 shows Lev's notes on whether the phrase 'live' should be used in the interface that he and his team are working on. Note that for confidentiality reasons, this is only a brief and redacted excerpt of a much longer copy doc. The text in Extract 2.3 was originally formatted as bullet points, which I removed for ease of reading; italicized parts were formatted in italics in the original text.

**Extract 2.3: Excerpt from copy doc**

1. What considerations are responsible for the current wording? Seems like we're
2. trying to convey two things: that this action makes something accessible to users
3. and it changes its status in our CMS.
4. We should reflect the way people currently talk about this but also make it easier
5. for new people to learn the tool with descriptive and clear labels.
6. *Live* isn't ideal because *live* could be misread as a verb and not an adjective.
7. Whether *live* is capitalized depends on if we're using the term generically or if
8. we're referencing a specific UI string.

Lev's notes are clearly oriented towards a dialogue or discussion with his colleagues, as the use of direct questions (line 1) and the frequent use of 'we' (lines 1, 4, 7, and 8) indicate. At the same time, however, these notes also showcase Lev's linguistic insight. Across this short explanatory note, Lev addresses a variety of linguistic and communicative matters, such as grammar (e.g. whether 'live' is a verb or an adjective; line 6), audience design (e.g. that the text should address what current as well as future users might expect; lines 4–5), and, more broadly, the communicative function of the text itself (e.g. what it is supposed to 'convey' to users; lines 1–3). To be clear, Lev's linguistic insight here might not necessarily be considered 'expert' knowledge by some sociocultural linguists, who may find observations such as whether a word is a noun or a verb relatively trivial. Still, I repeatedly observed Lev's colleagues recognizing comments such as the ones in Extract 2.3 as instances of linguistic expertise; it is this local recognition of expert knowledge that

[7] A wiki is a website with multiple interlinked pages that can be edited by users. Wikis are commonly used for knowledge management and documentation in software companies.

**Copy doc title**

*Description of the project/task, including, for instance: background/contextual information; previous discussions and decisions; links to related documents, web pages, and design files; or screenshots of the interface itself. Such descriptions can be brief, but they can also span several paragraphs.*

| Description | Initial content | Recommended content | Explorations | Notes |
|---|---|---|---|---|
| *Brief description of the purpose of the microcopy, e.g. "no results message"* | *The previous version of the microcopy* | *The new microcopy that the UX writer has written* | *Other, possible drafts of the microcopy* | *The UX writer's notes about the microcopy, which might include explanations, questions, notes about whether the text has been approved by others on the team, screenshots etc.* |
| *Brief description of the purpose of the microcopy, e.g. "no results message"* | *The previous version of the microcopy* | *The new microcopy that the UX writer has written* | *Other, possible drafts of the microcopy* | *The UX writer's notes about the microcopy, which might include explanations, questions, notes about whether the text has been approved by others on the team, screenshots etc.* |
| *Brief description of the purpose of the microcopy, e.g. "no results message"* | *The previous version of the microcopy* | *The new microcopy that the UX writer has written* | *Other, possible drafts of the microcopy* | *The UX writer's notes about the microcopy, which might include explanations, questions, notes about whether the text has been approved by others on the team, screenshots etc.* |
| *Brief description of the purpose of the microcopy, e.g. "no results message"* | *The previous version of the microcopy* | *The new microcopy that the UX writer has written* | *Other, possible drafts of the microcopy* | *The UX writer's notes about the microcopy, which might include explanations, questions, notes about whether the text has been approved by others on the team, screenshots etc.* |
| *etc.* | *etc.* | *etc.* | *etc.* | *etc.* |

**Figure 2.2** Schematic representation of a copy doc. Typically, this would be either a rich text document shared via co-editing services such as Google Docs or Dropbox paper or a wiki page

I refer to when speaking of linguistic expertise. In this way, Lev's notes not only serve to surface possible points of discussion with his colleagues, but they also constitute a strategic representation of his (meta)linguistic knowledge.

That Lev's work contains such metalinguistic reflection is not necessarily surprising. As Koller (2017) suggests, critical language awareness – and the ability to use and verbalize such knowledge in one's work – is crucial for contemporary language workers. Indeed, in my follow-up interview with Lev, he suggested that these copy docs helped him establish his position as a (language) expert in the team. A copy doc 'serves to rearticulate your role as neither someone who reviews somebody else's writing, nor somebody that writes something from scratch ... but as someone who's, like, kind of a co-designer', he explained. The reference to 'designer' (rather than writer) in Lev's explanation is certainly worth noting here. I will comment on this labelling practice in more depth in Chapter 3; for now, suffice it to say that in software companies, designers are typically valued more; that is, they are often paid more than UX writers and tend to have more influence. The point that Lev makes here, then, is that copy docs allow him to establish not just the nature of his expertise, but also that this expertise has symbolic capital further afield as well.

The metalinguistic comments in copy docs also illustrate nicely how UX writers not only put a lot of thought into the words they write, but that in doing so, they also construct implicit and explicit rules about the role and function of language in digital media interfaces. In other words, they engage in theorizing how language works. Evidently, their theorizing is not the same as that of sociocultural linguists; what I mean is simply that these wordsmiths also produce explicit formulations of how language works. In this sense, UX writers are not just language workers but also language experts (cf. Thurlow, 2018; Thurlow & Britain, 2020), regardless of whether their theorizing takes the same form – or perhaps nuance – as that of sociocultural linguists. In light of this, I want briefly to tease out two aspects that UX writers seem to understand particularly well about language in digital media interfaces: its dynamism and malleability, as well as its role in shaping users' subsequent (inter)actions.

### 2.5.2 *Interface Texts as Dynamic and Malleable*

The UX writers I talked to at Novus tended to understand their writing as temporary and provisional; often, there was neither a clear origin nor end point to it. As such, they understood the microcopy they wrote as 'an ever-changing text' (cf. Hicks & Perrin, 2014, p. 233, about Wikipedia), as something that was constantly open to revision and change. This provisionality and openness of the text starts, quite simply, with the fact that these people's text production rarely begins with the creation of something entirely new, but instead involves revising, (re-)assembling, and re-writing other previously written texts. Consider Extract 2.4, where Lev talks about how he usually writes microcopy.

**Extract 2.4: Work process interview with Lev**

1. A lot of the work that I'm doing is not writing stuff from scratch. . . . I usually work
2. from a designer's recommended content. But within that context, about 10% of
3. the time, it's like finding stuff that I had written previously to sort of cannibalize it.

As Lev says, most of his writing does not involve creating something 'from scratch' (line 1); rather, his text production usually starts with text that has been written by someone else (lines 1–2). Other UX writers from Novus similarly explained that their work often started with placeholder text written by non-UX writers, which they would subsequently revise through their work. UX writing – like many other kinds of professional writing – thus involves multiple people shaping the text as intermediary co-authors (cf. Schindler & Wolfe, 2014). While the UX writer is typically seen as the person responsible for the final text, these other temporary co-authors can nonetheless have a significant impact on the microcopy, especially if they hold considerable institutional power, such as the company's legal team or highly ranked executives. This is, perhaps, why copy docs (cf. Figure 2.2) are so important: they allow UX writers to not just record decisions but also establish their expertise vis-à-vis this multiple authorship.

In addition to the multiplicity of authorship, Lev's writing process also entails a multiplicity of texts. In this regard, he not only revises and re-assembles texts that someone else wrote but also re-assembles his own writing (lines 2–3). Lev speaks of 'cannbaliz[ing]' (line 3) previously written texts, implying an approach whereby the existing text is not preserved in its entirety, but rather disassembled into parts that can be reused and parts that are discarded. There is a non-determinacy to the texts he writes. In this way, the work of UX writers seems to be best characterized as a shifting assemblage of the kind that posthumanist scholars tend to describe (see, e.g., Pietikäinen, 2021 for a sociolinguistic introduction to assemblage theory). In fact, later in our interview, Lev described his work as an act of 'putting together unsteady assemblages', emphasizing again the temporary and unstable nature of these texts. In any case, the microcopy Lev writes seems to be something that is *re*-written rather than written; it is dynamically (re)assembled from existing texts to create something new – always under the premise that this new text, too, might at some point be disassembled and re-assembled.

The rewriting and reuse of existing texts is of course not unique to UX writing. As mentioned earlier, technical writing is similarly characterized by distributed forms of writing, whereby texts are frequently created through a process of reuse and recycling (e.g. Slattery, 2007; Swarts, 2010). Other kinds of writing are also centrally concerned with (re)writing texts based on existing sources. News articles constitute a well-researched example in this regard (see Catenaccio et al., 2011); another obvious example is academic writing itself, where citing practices might be understood as a specific form of rewriting. Nonetheless, something that is particularly intriguing and perhaps different about UX writing is that even once the interface text is published or released, it is still open to further instances of rewriting. As Lev put it '[micro]copy is pretty malleable'. It can and often will be changed, even if it at some point has been considered 'done'. This shows that UX writers like Lev are particularly attuned to the fact that unlike print-based texts, digital texts – and interface texts in particular – are never really fixed but always subject to potential change. They are characterized by a certain provisionality, making them a dynamic and malleable phenomenon rather than a static and bounded object.

In contrast to UX writers' understanding of texts as dynamic and malleable, discourse scholars still often tend to focus on finished products rather than textual productions (cf. Macgilchrist, 2017). As Macgilchrist points out in her work, though, it is often the processes behind these otherwise finished products – in other words, their text trajectories – that can be especially insightful for understanding how and why particular social discourses become embedded in specific texts (e.g. Macgilchrist & van Hout, 2011; Macgilchrist et al., 2017). In this way, the applied expertise of UX writers and their intuitive understanding of the dynamism and malleability of texts also highlight some of the shortcomings of certain scholarly traditions in sociocultural linguistics.

### 2.5.3 *Interface Texts and/as Action*

UX writers also appear to understand interface texts as dynamic in a different sense, namely in that they often approach these texts not as something that is meant to be *read* but as something that is meant to be *acted upon*. Sometimes, this is made quite explicit. For instance, one of my interviewees at Novus, Christina, explained that for the particular project she was showing me (a sign-up interface), she had made sure to use 'aspirational language' in order to get people to register for a paid subscription. 'It's about enticing people do something they might not do otherwise', she explained. For confidentiality reasons, I cannot discuss the actual microcopy Christina wrote; nonetheless, her explanation that it was about 'enticing people' suggests that she understands her writing as concerned with gauging – and influencing – the subsequent actions of users.

To be clear, not all UX writers were as explicit about the anticipated or designed consequences of their work. In general, however, this orientation towards the action potential of one's writing was also evident in my discussions with other UX writers at Novus. As a way of illustrating this, I offer Extract 2.5, where UX writer Diana talks about the microcopy she wrote for introducing users to a new software functionality.

**Extract 2.5: Work process interview with Diana**

1. We're trying to get [the users] through this thing. So, "introduction" means low
2. complexity, and so immediately, their brain goes "Okay, I can do this pretty
3. quickly." And they're gonna see that it's a very short box, and so that's going to
4. help them engage. . . . Here's a single effort, and you get a lot out of it, so do that!

Not surprisingly, Diana's main objective with this interface text is not necessarily to get users to read what she wrote but to ensure that they can navigate through the interface quickly; the point is, as she says, 'to get [the users] through this thing' (line 1). Here, UX writing shows similarities to advertising copywriting, which, similarly, is not about getting people to read a text but rather to act on it in a way that is favourable to the advertiser, perhaps without even thinking about the text itself (O'Shaugnessy & O'Shaughnessy, 2003; see also Droz-dit-Busset, forthcoming). Diana mentions two techniques she uses for this: (a) choosing words that are low in complexity (lines 1–2) and (b) making sure that the text looks 'very short' (line 3). By attending to the verbal and visual (or material) aspects of her writing, Diana strives to reduce the actions users have to take into 'a single effort' (line 4). This reduction of the number of actions users can take in response to an interface was also mentioned by some of the other interviewees; it is in this way that UX writers do not only attend to action but also effectively design and pre-define particular (inter)actions for users.

Diana's orientation to the (imagined) actions of users also becomes apparent in how she describes her work more generally. While talking through her writing choices, she voices what users (might) think – 'I can do this pretty quickly' (lines 2–3) – and even produces an imperative that directly addresses hypothetical users: 'you get a lot out of it, so do that!' (line 4). Both of these statements imply action: the latter demands that users 'do' whatever the interface asks them to, while Diana's hypothetical verbalization of users' thoughts envisages that they have already agreed to act upon the interfaces' proposition. Again, what emerges here is that Diana approaches microcopy less in terms of what it says and more so in terms of what it does – or, more specifically, in terms of what it makes users do. Understood thus, the interface text inevitably becomes implicated in action.

That texts are not just static entities but resources people use to take action is certainly not a new insight. In fact, scholars of mediated discourse analysis have long argued that there is value in looking at discourse through the lens of (mediated) action, putting not text or discourse at the centre of analysis but rather the social actions people take – with or without texts (cf. Norris & Jones, 2005). As such, interface texts might be understood as something akin to what Norris (2004) calls a 'frozen action' – an action that is materialized or embedded in an object. Importantly, though, what UX writers like Diana focus on are not the actions they themselves take with this text but rather the subsequent actions of others; in this way, they seem to be (implicitly, at least) aware that their work is implicated in an agential chain (cf. Latour, 1984) that connects producers and users through the software interface. This of course has to do with the particular, structural conditions of their work; nonetheless, they understand – perhaps better than sociocultural linguists – that interface texts may also pre-scribe the subsequent actions of users. Owing to this, the work of UX writers ultimately has considerable normative impact by shaping how users (can) interact with digital media in the first place.

## 2.6 Conclusion: Designing Words

My aim in this chapter has been to offer a broad introduction to the language work of UX writers, with a specific view to both the profession as such as well as the linguistic expertise of these language workers. UX writers are not just a typical example of contemporary language workers, they are also a perfect illustration of Thurlow's (2020c) 'wordsmiths': their work hinges on its taking place behind the scenes – indeed, the hallmark of a good UX writer is that their work becomes unnoticeable to users – and yet, the words these professionals produce are influential not just in determining what people can do with and through digital media but also in shaping cultural norms (cf. Massanari, 2010). Additionally, as I have discussed, UX writing is highly valued, both in symbolic and material terms. In the relative hierarchy of language work, these are certainly more privileged or elite language workers.

Nonetheless, it is worth noting that while UX writing may be highly valued, within the software design industry itself, UX writers (along with other kinds of language workers) are generally paid less than the people working alongside them, such as project managers, designers, and especially software engineers. They are also employed at a much lower rate, and they tend to have less influence and authority when it comes to making important decisions. Indeed, as one of my interviewees explained, despite the rise in popularity that UX writing has seen in the last decade, the reality is that most of the microcopy ordinary users encounter when using digital media has not been written by a UX writer but by someone – anyone – else: a designer perhaps, or an engineer, or a project manager – whoever happened to have time to write some quick words.

When UX writers are involved in writing interface texts, they are often under constant pressure to prove their value vis-à-vis other professional disciplines and to show that their 'designing' of words is as valuable as the design work of their colleagues. In light of this, it is perhaps not surprising that UX writing is also a much more woman-serviced profession than, for instance, engineering (or even design). In fact, two-thirds of my interviewees (16 out of

21) were women. I don't mean to suggest that this number is necessarily representative, but it is certainly not surprising. Even in the software and technology industry, communication seems to be predominantly – or prototypically – women's work (cf. Cameron, 2000b). Some of my interviewees sexistly attributed this to women having more empathy and therefore being better suited for this kind of work, while others saw it as a result of structural pre-conditions, such as women being groomed from a young age towards language-oriented tasks and skills like reading stories and learning foreign languages. In any case, the gendered nature of the work surely influences the particular status UX writers (can) have within their professional field.

To complicate things further, the linguistic work UX writers do is often not recognized by their colleagues: 'the word magic comes up a lot', one of my interviewees explained. While this may sound flattering, it also constitutes a devaluation of UX writing. As she went on to explain, for her, 'there's a love/hate relationship with that word, because it's not magic – that's 15 years of experience'. Here, then, one sees the crux of UX writers' status as wordsmiths: what they do may be seen as impressive but seemingly inexplicable, and while their work may be admired, they are often not given the same status as other experts. The expertise of UX writers tends to be based on experience rather than a clear educational path and the 'hard' fact of a specific degree; others may thus (unjustly) equate it with innate talent rather than hard work (cf. Duchêne, 2020 on the naturalization of linguistic labour in general). Thorough this, UX writers' expertise can be (and often is) treated as lesser than that of their colleagues in software engineering, UX research, or even design.

In spite, or perhaps precisely because, of their reliance on experience rather than taught rules, I suggest that UX writers hold a valuable perspective on language that makes it worth examining not just their professional practice but also the way they understand and theorize language for themselves. As I have shown, the day-to-day work of UX writers in fact depends on their ability to articulate and demonstrate metalinguistic knowledge (again, see Koller, 2017, who argues that this is central to language work in general). In this regard, I have identified how UX writers understand interface texts as dynamic rather than static in two distinct but related ways. On the one hand, UX writers seem to be particularly aware of the dynamic and malleable nature of interface texts and of how interface texts – but perhaps also digital texts more generally – might be best understood not as something fixed but as flexible and constantly open to change. Second, I have shown that UX writers tend to emphasize action in their text production; this, too, points to the dynamic nature of UX writers' text production in so far as they tend to understand interfaces not as texts to be read but rather as (social) action that changes what users can, should, or will do. Both of these aspects point to a posthumanist understanding or approach to language, which is something I will return to at various points throughout this book.

Ultimately, the initial insights into UX writing outlined in this chapter indicate that UX writers hold considerable cultural-political power as the professionals shaping key aspects of the language of digital media interfaces. Importantly, this is often enmeshed with their own understanding of how language works or should work. Overall, then, it is worth examining more closely not just how UX writers use language in their work but also how they themselves approach language and how these understandings of language depend on the material and structural conditions of their work. In Part II of my book, I therefore turn to examining UX writers' reflexive understanding of their profession and their work more closely. Concretely, Chapter 3 addresses how UX writers articulate their linguistic expertise as a unique skill and how this discursive construction of skills allows them to professionalize and legitimize their own work. In Chapter 4, I then turn to one particular aspect of UX writers' work: (non-)creativity. In this regard, I discuss how UX writers understand their own work as creative (or not), arguing that they frequently prioritize the discursive function of interface texts over their verbal form. Across these two chapters, I am particularly interested in how UX writers ultimately claim status and expertise – in other words, how they work to establish themselves not just as language workers but also as (relatively) elite wordsmiths.

*II*

# Establishing Status

*UX Writing as Elite Language Work*

# 3 Skilling the Writer

## *UX Writing and the Professionalization of (Elite) Language Work*

While language work has been a key interest in critical sociolinguistics for some time, much of this research has focused on low-status language work such as work in call centres (e.g. Cameron, 2000b; Boutet, 2012) or in the tourism industry (e.g. Heller et al., 2014). Often, such research is undertaken with a loosely Marxist view vis-à-vis commodification and issues of exploitation and alienation (see Block, 2018 for an overview). As I have established in Chapter 2, UX writing does not fit this bill. As part of the so-called tech industry (cf. Dorschel, 2022), UX writers can earn a relatively high salary; at the same time, the UX writers I encountered through my research were all proud of their work and expressed strong personal identification with what they were doing. As one of my interviewees explained, 'a lot of people in this profession, myself included, sort of define themselves by this role and by what they're doing.' Given this, it seems inadequate – or at least insufficient – to examine the work of UX writers through the analytic lens of commodification and exploitation. This is not to suggest that UX writing is not enmeshed in the wider hierarchies of language work; certainly, the privileged status of UX writers is rooted exactly in people profiling themselves competitively and claiming higher-status symbolic capital for economic conversion. My point, however, is that this happens not just through commodification but also – or especially – through discursive processes of professionalization and skilling, both of which, I suggest, ultimately function as status-producing mechanisms.

In this case, I am specifically interested in how UX writers as members of a relatively new and emerging professional group discursively legitimize and professionalize their language work. For example, it is interesting – and perhaps only a slight coincidence – that the interviewee I quoted earlier refers to UX writing as 'this profession', rather than, say, 'this job', 'this kind of work', or 'this occupation'. While I myself am also variously referring to UX writers as professionals, this notion is one I mean to problematize in the current chapter. As a label, 'professional' has a certain cachet; professionals are workers who typically have not only expert knowledge but also higher social status and, depending on how one defines the term, also a specific legal status (see Adams, 2010). Having said that, who counts as a professional and what counts as a profession is far from clear-cut. Indeed, as Abbott (1988) points out, it is often the fact that something is called a profession that makes it one, even in scholarly literature. Both in theory and in practice, then, profession is a discursively negotiated category, whereby some people are warranted as professionals while others are not. It is this intersection of discourse, work, and status that interests me in the present chapter, where I examine the discursive tactics UX writers use to professionalize their job and thereby negotiate the status of their work.

In what follows, I first offer a brief overview of critical sociolinguistic research on language work, specifically with a view to critiques about the field's focus on language commodification and its – by contrast – more modest engagement with skilling as an analytic concept. I then turn to discussing the discursive construction of skills and professions as the theoretical framing for my analysis. In this regard, I orient to research on professional discourse as well as scholarship in the sociology of professions to outline what I understand by profession and professionalization. Building on this, and as the main part of this chapter, I then examine how UX writers discursively legitimize and professionalize their own work. For this, I draw on a variety of complementary sources to explore professionalization in UX writing; this includes participant observation at a training course, interviews with UX writers, and written texts such as books about UX writing. In my analysis, I observe the construction, codification, and indexing of 'writing-as-designing' as a unique skill in UX writing and discuss the various rhetorical tactics UX writers deploy to this end. Ultimately, I connect this case study to broader questions of language work, suggesting that in order to understand not just the elite language work of UX writers but also hierarchies of language work more generally, it can be fruitful to broaden such scholarship with a view to professionalization and skilling.

## 3.1 Language as Commodity, Capital, and Skill

Critical sociolinguistic research on language work typically concerns itself with how language has become increasingly commodified in the wake of late capitalism (e.g. Duchêne & Heller, 2012). Whether language can be seen as a true commodity has, however, been criticized by a number of researchers (e.g. McGill, 2013; Block, 2018; Simpson & O'Regan, 2018; for a recent overview, see the edited collection by Petrovic & Yazan, 2021b). Many of these scholars point to a lack of engagement in critical sociolinguistics with Marxism specifically and economics more generally. Indeed, Block (2018) goes so far as to suggest that language commodification has become a buzzword scholars use to affiliate themselves with a particular research tradition without properly engaging with the concept of commodification itself.

Scholars subscribing to an orthodox Marxist approach thus take issue with how 'commodity' is understood in much literature on language work and with the lack of precision in how it is theorized. Petrovic and Yazan (2021a), for instance, note a tendency for conflating the terms commodity, capital, instrument, and resource. Similarly, Simpson and O'Regan (2018) criticize the tendencies of critical sociolinguists to 'abstract away' the idea that value is produced through labour (p. 159). Both take issue with the fact that the label 'commodity' is bestowed onto something that has not been produced through labour. For Heller and Duchêne (2016), the main reason for disagreement between different research traditions lies in how language as such is conceptualized. Indeed, while Petrovic and Yazan (2021a) seem to understand language as a natural phenomenon that is 'rooted . . . also in our physical nature' (p. 27), Heller and Duchêne (2016) take it to be 'a social practice that consists of circulating communicative resources' (p. 139). Evidently, these two positions differ in their understanding of the ontological status of language. Additionally, they differ in their conceptualization of commodity: one employs a Marxist understanding of commodity (see Marx, 1867/2019), while the other orients more to Bourdieu's (1977) notion of the linguistic marketplace.[1] Given my interest in issues of prestige and status in language work, my own work naturally aligns with the kind of research Duchêne and Heller pursue. I thus use the term capital (rather than commodity) to indicate my theoretical orientation to Bourdieu and his work on linguistic capital. Capital in this sense is connected to work, but it is not necessarily a product of labour, rather, it is a result of social power relations.

While scholars in Marxist-styled and Bourdieusian-styled scholarship disagree on the status of language as a commodity, they tend to agree that language as a *skill* can have economic value. In Marxist terms, it is 'the selling of one's labour power with an increased [linguistic] capacity' (Petrovic & Yazan, 2021a, p. 27) that is involved in economic exchanges. Following Bourdieu (1977), meanwhile, language skills can be understood as linguistic capital that may be transformed into other kinds of capital (including economic capital). This second position aligns with Urciuoli's (2008) observation of the rise of a 'skills discourse' in late capitalism, whereby workers are expected to see themselves as 'bundles of skills' based on which their value in the labour market can be determined. For Urciuoli (2008), language skills therefore 'acquire value to the extent that they add value to a worker's value in the labor market' (p. 472; see also Block, 2021); it is in this way, and following

[1] Thus, Simpson and O'Regan (2018) shift the debate to problematizing the ontological status of commodity rather than language, suggesting that Heller and Duchêne fail to address the material labour conditions that turn something into a commodity. They, too, follow a Marxist rather than Bourdieusian tradition.

Shankar and Cavanaugh (2012), that one might think of language skills as an objectification of language.

Evidently, skills and skill discourse is not a new topic in research on language work (e.g. Fairclough, 1999; Cameron, 2000a; Urciuoli, 2008). In this case, I orient primarily to Urciuoli's (2008) work on the neoliberal conceptualization of skill in contemporary workplaces. To her, skill discourse is about how the notion of skill is conceptualized and fetishized in late capitalism, where skill terms such as *communication, leadership*, and *teamwork* are recast from social act into 'soft skills' (see also Cameron, 2000a). Importantly, Urciuoli (2008) observes that it is the individual that is responsible for acquiring these skills (see Kraft, 2020 for a recent case study in this regard). Urciouli's work further points to the elusive nature of skills. She suggests that skill-related terms are strategically deployable shifters, words that are denotationally vague and whose meaning depends on the context in which they are used. Owing to this, a term like *communication skills* can be deployed to indicate alignment with particular values or discourses without giving any specific sense of what one actually means by the term.

Research on migrant workers equally indicates how vague a term such as *language skills* is, illustrating perfectly that whether speaking a certain language counts as a skill is never clear cut and often influenced by issues of race and gender as much as by people's actual linguistic/communicative competence (e.g. Allan, 2013; Flubacher et al., 2016). Above all, this kind of research shows that the notion of skill is not a descriptive term but rather an attribute that is negotiated in and through language. Skills and skilling therefore offer a particularly interesting angle for exploring language work. Specifically, I suggest that a focus on the discursive construction of skill allows scholars to better understand hierarchies of language work – how it is that the linguistic labour of someone working in a call centre or tourism agency is generally valued less than that of a UX writer working at a software company, and how it is that some language workers are regarded as so-called professionals while others are not.

## 3.2 The Discursive Construction of Skills and Professions

Despite – or perhaps because of – their elusive and highly discursive nature, skill ideologies take a prominent role in how people construct professional boundaries and professional status. Beliefs about what counts as a skill and who might legitimately claim that skill can be what sociologists call a social closure strategy (Parkin, 1979; cf. Weber, 1978), that is, a means by which social groups close off access to resources for other groups – in this case, a means for one professional group to exclude others from particular knowledge and professional opportunities (see Adams, 2018). In this vein, as Steiger

(1993) argues, skill is socially constructed in two senses. On the one hand, there is a social construction of skill itself, that is, the process determining which skills are indeed required for a specific job. On the other hand, the construction of skilled status, that is, the social construction of groups said to possess these required skills, is also a matter of discourse. Steiger (1993) thus suggests analyzing skill not as a question of technical ability but as an ideological discourse; this, he argues, allows scholars to examine how skill discourses are entangled with issues of pay, power, and privilege.

Language and discourse take on a crucial role not just in the social construction of skills but also in the construction of what counts as a profession. In sociocultural linguistics, this has been explored in particular by scholars looking at professional discourse (e.g. Gunnarsson et al., 1997; Kong, 2014). Professional discourse in this sense is typically understood as the talk and/or texts produced by professionals in professional contexts; this includes interactions between professionals of the same or different domains as well as interactions between professionals and lay people. Often, the term professional is broadly conceived; Gunnarsson (2009), for instance, uses the term 'as a synonym for "paid-work related"' (p. 5), while Kong (2014) applies it to 'any individuals who have undergone specialist training or education before they are eligible to become members of a certain profession' (p. 2). Understood this way, professional discourse is a homogenous category encompassing a variety of discursive phenomena. What I am particularly interested in, however, is the role that professional discourse plays in socializing people into professional roles (cf. Kong, 2014, chapter 2) and the connection between expert knowledge and the language people use to construct such expert status.

In general, professional status is closely linked to language use. Speaking a particular register, for instance, can be fundamental to being seen as part of a particular professional community. In this way, Porcello (2004) shows how learning to be a sound engineer entails learning to speak like one. This professional register, says Porcello, can in turn function as proof of one's professional training and experience. Knowledge of linguistic practices then serves both as a criterion for membership in the professional community, and it can constitute the basis for internal stratification. Especially for professions that entail more direct contact with lay people (e.g. clients, patients, or customers), professional status is also established through one's position vis-à-vis others. Examining how social workers claim professional expertise, Hall et al. (1999) therefore suggest that a professional's position of expertise is established through a process of role differentiation, whereby professional identities are legitimated in interaction vis-à-vis clients through, for instance, the construction of particular client types that serve as a foil for the role of the professional.

As a whole, though, literature on professional discourse is typically more concerned with language produced *by* professionals rather than language produced *about* professions. It is for this reason that I also turn to research on the sociology of professions to understand how professions are socially and discursively constructed. While there is no consensus in the field over how to define professions, scholars have generally understood professions as closely linked to the concept of skill, with professions being conceptualized, loosely put, as 'an occupational group with some special skill' (Abbott, 1988, p. 7). Abbott (1988) thus argues that what defines professions is that they develop 'a knowledge system governed by abstractions' (p. 9). In other words, professions are held together by a codification of the skills (including language skills) that are said to be necessary for and characteristic of their work. This, too, is not a neutral exercise; establishing and maintaining disciplinary knowledge is also a matter of competition.

Emphasizing the competitive aspects of professions, Bourdieu (Bourdieu & Wacquant, 1989) views a profession as a field, structured not just by internal power relations but also by the position of the field vis-à-vis other fields. Understood thus, the limits of a profession are 'always at stake in the field', and the members of the profession have to 'constantly work to differentiate themselves from their closest rivals' (Bourdieu & Wacquant, 1989, p. 39). It is for this reason precisely that Bourdieu calls for an 'undoing' of the social construction of profession, suggesting that profession ought to be treated as a folk concept (see also Freidson, 1983, who argues the same point). This shifts the question from whether something is a profession to when and how people attribute professional status to something. For Freidson (1983), this entails examining how 'people in a society determine who is a professional and who is not, how they "make" or "accomplish" professions by their activities, and what the consequences are for the way in which they see themselves and perform their work' (p. 20). Turning now to UX writing, I align myself with Bourdieu and Freidson, seeking to interrogate the status claims that UX writers advance to distinguish their own work internally and vis-à-vis other fields of language work. I understand such claims as a rhetoric of professionalization; understanding professionalization itself as 'those processes that combine to raise an occupation's professional status, or maintain its status' (Adams, 2018, p. 21). In other words, it is through claims of professionalization that people negotiate the prestige, status, and power of their work.

## 3.3 The Data

Questions of professionalization and skilling are inevitably bound up with questions of how people are schooled into professional practices. Indeed, while scholars in the sociology of professions may disagree on the exact definitions of 'profession', they typically agree that knowledge consolidation is an important aspect of professionalization (Adams, 2018; see also Abbott, 1988). A good example of this is schools and degree programmes. In light of the weight education is given in this literature – and bearing in mind the complex influence that formal education can have on professional development in general (e.g. Nyström, 2009) – this chapter focuses on processes of skilling and professionalization by examining how UX writing is formally taught and learnt.

As noted in Chapter 2, 'UX writer' is a relatively recent job title – even those of my interviewees who claimed they had been doing this kind of work for over a decade said that the title itself had appeared only in the mid-2010s. Almost all of my interviewees claimed to be self-taught, having acquired the necessary skills through independently reading books, attending conference talks, and informal learning from more senior colleagues. Given the heterogeneous approach to skilling that my interviewees reported, I draw on a variety of sources to examine the construction, transmission, and codification of professional knowledge in UX writing. This includes much of the material I discussed in my mapping of the profession (see Chapter 2); specifically, I return to my informational interviews and to some of the books on UX writing I analyzed. Primarily, however, this chapter emerges from fieldwork I conducted at a three-day online course on UX writing, which I attended as a participant observer.

Since the late 2010s, a variety of workshops and courses on UX writing have started to appear along with the new job title. Most of these courses are taught by industry practitioners, who engage in this as a side activity next to their day-to-day work as UX writers. Such training is typically offered as half-day workshops at industry conferences, as self-paced online courses, or as online classes spread out over the course of a few days or weeks. Most courses and workshops are organized through private sector companies and do not claim any academic or professional accreditation, although there are a few recent programmes that are affiliated with professional schools and universities.

I conducted my fieldwork in November 2021, at a time when all available courses were offered as online courses due to the ongoing COVID-19 pandemic. (Some well-known courses on UX writing had been offered as online courses before that, too.) This made choosing a particular course for fieldwork both easier and more difficult. On the upside, it gave me the opportunity to easily attend any course I could find, even if it was offered halfway across the globe. On the downside, it meant that possibilities for interacting with other participants and the instructors during breaks or after class would be much more limited. These constraints impacted the course I chose for my fieldwork; regardless of these advantages and disadvantages, though, participating in an online course certainly mirrored the experience of anyone who was looking to take a course on UX writing at that time.

The course I chose for my fieldwork was a three-day, live online course that promised hands-on learning and interaction with other participants due to a small group size of approximately twenty participants. The course itself was taught by two established UX writers, Tom and Steve, both of whom were based in the United States.[2] This reflects the reality of the field (most UX writing courses I found were taught by US-based professionals); at the same time, it also allowed me to remain true to my pre-pandemic fieldwork plan, since I had originally planned on-site fieldwork in the San Francisco Bay Area as an epicentre of professional trends in the software design industry (and with this, the UX writing industry). It was a course I had previously seen mentioned in various UX writing newsletters and in social media posts, which I took as an additional sign of the popularity of the course. I attended the course as a full participant, taking part in all activities in order to capture the real, lived experience of learning about UX writing. The instructors and the other participants were aware of my status as a researcher; however, to respect the privacy of all attendees, I did not record any interactions during the course. For the current analysis, I therefore draw on my fieldnotes from the course and on two one-hour interviews I conducted with the two instructors one month later.

To analyze the data from the UX writing course, I started with revisiting my fieldnotes in order to identify points of interest: aspects that either surfaced often and/or that the instructors had especially emphasized. One topic that stood out was the emphasis the instructors had put on writing-as-designing as a particular, perhaps distinctive, skill characterizing UX writers (more on this below). With this in mind, I listened to the two interviews with the instructors and noted down instances where writing-as-designing was mentioned in order to subsequently transcribe these passages and compare them with my fieldnotes.

As mentioned above, my analysis is also informed by my profession mapping and, especially, by my close reading of three trade books on UX writing (cf. Chapter 2):

- *Writing Is Designing* by Michael J. Metts and Andy Welfle (2020)
- *Strategic Writing for UX* by Torrey Podmajersky (2019)
- *Writing for Designers* by Scott Kubie (2018)

[2] These names are pseudonyms. At the time of writing, the UX writing community is relatively small, and there are only a handful of well-known courses. In light of this, I do not offer more information about the instructors, and I offer only limited information about the course itself in order to keep the identity of the two instructors as anonymous as possible.

Here, too, I was specifically interested in where and how writing-as-designing surfaced. To this end, I re-read my previous notes on the books in order to identify any passages where particular (writing) skills were discussed. In the analysis that follows, I take the UX writing course as a starting point but trace connections to these other texts and to interactions I had with UX writers when I conducted initial informational interviews (again, see Chapter 2 for details). I understand these connections as pointing to an ongoing construction and codification of skills within the domain of UX writing, which is part and parcel of the professionalization of UX writing as a whole.

## 3.4 Analysis: Professionalizing UX Writing

In what follows, I offer an overview of various ways in which UX writing is discursively constructed as distinct and legitimate work. Following research in professional discourse studies and the sociology of professions, I trace the construction of 'writing-as-designing' as a unique skill that UX writers claim for themselves and discuss the construction, codification, and indexing of this skill as key moments in the professionalization of UX writing. My analysis begins, then, with an examination of how such knowledge is consolidated and framed as specific to UX writing.

### *3.4.1 Writing-as-Designing: Constructing and Managing Professional Skills*

As part of my research, I was particularly interested in how UX writers are socialized into specific professional practices and how they learn specific skills deemed necessary for (and perhaps unique to) UX writing. In this regard, I found that there was one skill that surfaced time and again in my research: design. Design itself is a complex term with multiple meanings, encompassing both aspects of art (e.g. fashion design) and engineering (e.g. industrial design). Vis-à-vis art, design is typically understood as more purposeful; at the same time, and in contrast to engineering, design is usually said to (also) orient to aesthetics (cf. Walker, 2009). In the context of UX writing, my interviewees typically used the term to refer to UX design (cf. Chapter 2) or sometimes more narrowly to refer to the visual design of software interfaces.

As discussed in Chapter 2, many of the prominent books on UX writing explicitly frame UX writing as a design activity – recall, for instance, the telling book title *Writing Is Designing* (Metts & Welfle, 2020). While I was not surprised by this – my own experience of working in UX taught me how highly regarded the label 'design' could be in this environment – I was still somewhat struck by how persistently 'writing' was eclipsed by 'designing' in a profession that quite literally had the word 'writing' in its name. In line with this, the UX writing course I attended started with a differentiation between how writers and designers would approach creating interface texts. Writers, our instructors explained, might think mostly about what the right words are, focusing on vocabulary, grammar, and orthography. By contrast, a designer, we were told, starts with inquiring about the context of a specific interface text, considering, for instance, what happens before or after the text appears on a user's screen or how specific words would affect a user's behaviour. Both of these approaches, 'writing-as-a-writer' and 'writing-as-a-designer', appear to lead to the same result, namely written words. However, there seems to be a difference in how writing as such is conceptualized in each instance.

As scholars in sociocultural linguistics show, writing is both a complex semiotic object and a complex semiotic practice (e.g. Blommaert, 2013). In this vein, Lillis (2013) proposes that writing has a variety of modal dimensions that influence how people use writing. Specifically, Lillis discusses six dimensions of writing: writing as inscription, writing as verbal, writing as material, writing as technologies, writing as visual, and writing as spatial. Two of these seem particularly relevant when it comes to UX writers: writing as verbal and writing as technology.[3] Writing as verbal concerns, as the name suggests, how writing 'has been defined in terms of its verbal nature' (Lillis, 2013, p. 26); this includes things such as sentence- and text-level grammar as well as rhetoric or stylistic choices. In simple (or simplified) terms, one could say that writing as verbal is about choosing specific words. This aligns with what my UX writing instructors framed as a writer's approach to writing interface texts: one that is primarily focused on lexical choice.

Understanding writing as a technology, on the other hand, foregrounds that writing is typically also used to *do* things. Both as object and practice, writing therefore constitutes social action. In this regard, Lillis (2013, pp. 30–32) discusses how different ways of writing (e.g. writing on a Facebook wall vs. spraying graffiti on a physical wall) make use of different technologies (e.g. digital text production vs. a spray can) with different practical and social effects. Lillis's own discussion revolves around technology in a technical or material sense; nonetheless, I suggest that implied through this is also an understanding of writing itself as a technology or as a cultural tool (Vygotsky, 1981; Norris & Jones, 2005), a concept that scholars in mediated discourse analysis use to describe the technical but also psychological tools people use to carry out social actions. Understood this way, writing is

[3] Lillis (2013) herself uses the plural form for this ('writing as technologies'). Orienting more strongly to scholarship in mediated discourse analysis – which Lillis's discussion seems to align with but does not actually mention – I find the singular form ('writing as technology') more fitting. This second form makes the distinction more explicit between writing as a cultural tool and other (technical) tools that one might use *for* writing (such as a pen, a computer, or a spray can).

something people use to act upon the world. This conceptualization of writing overlaps with what the instructors of my UX writing course framed as a designer's approach to writing: a writing practice that is highly contextualized and primarily concerned with the effect of specific words. In this sense, these two different kinds of writing, writing-as-a-writer and writing-as-a-designer, nicely map onto Lillis's (2013) modal dimensions of writing as verbal and writing as technology.

My interviews with the two course instructors, Tom and Steve, confirmed the observations I had made based on my participation in the course itself. To this end, Extract 3.1 shows Tom's response to my question about the main aims of the course. I am particularly intrigued here by his reflexive understanding of the term writing.

**Extract 3.1: Writing-as-designing in UX writing course (Tom)**

1. We're aiming to teach a new way of looking at writing as a design method. So
2. writing– there are just so many different types of writing, and I think that makes
3. it different in a way from, if you were to say, design. . . . If you say "design," people
4. think of, like, "Oh, yeah, it's how it looks." But for writing, we're trying to teach
5. people, like, this isn't about creative expression, it isn't about prose, it isn't about
6. a clever line; this is about solving a problem for people using this system. They're
7. not interested in if you're clever or not; they're interested in getting through their
8. lives. And your success is measured by how well you help them do that.

As Tom explains, what he and Steve aim to teach people in their UX writing course is to 'look[] at writing as a design method' (line 1). Evidently, this is aligned with and reminiscent of what he and Steve explained about writing (and designing) in the course itself. Tom continues by pointing out the vagueness of the term writing ('there are just so many different types of writing', line 2), which he contrasts with the term design as a relatively unambiguous concept (lines 3–4): design, apparently, is 'how it looks' (line 4). As mentioned above, the term design itself has different indexicalities, too; in this case, however, Tom presumably refers to visual design, a professional field adjacent to UX writing (cf. Chapter 2). Tom's claim that this adjacent profession is somehow easier or simpler to delimit might be understood as a way of highlighting the distinct and (supposedly) more complex nature of UX writing as its own field.

While Tom does not explicitly elaborate on the 'many different types of writing' he alludes to, he does use the existence of other types of writing as an entry point for distinguishing UX writing. It is here that one can see parallels to Lillis's modal dimensions of writing. UX writing, Tom explains, is not about 'creative expression', 'prose', or writing 'a clever line' (lines 5–6), suggesting that in UX writing, the focus should not be on words for their own sake. Rather, Tom says, UX writing is 'about solving a problem for people' (line 6) and helping them 'get[] through their lives' (lines 7–8). This explanation again establishes a distinction between focusing on words (writing as verbal, or writing-as-a-writer) and focusing on action (writing as technology, or writing-as-a-designer).

At the same time, Tom's account indicates that the struggle over the term writing is linked to how writing can be claimed as a skill that is unique to UX writing. As a generic activity encompassing 'so many different types of writing' (line 2), writing is something many people do; but writing in a way that 'solv[es] a problem for people' (line 8) is more specific and, Tom seems to suggest, new or at least different from other kinds of writing: this is 'writing as a design method' (line 1). The struggle over what kind of writing UX writers do can thus be understood as a struggle over the limits of the professional field (cf. Bourdieu & Wacquant, 1989), whereby this 'new way of looking at writing as a design method' (line 1) becomes a skill that can be claimed as unique to UX writing. In this way, it becomes something that UX writers can mobilize in order to differentiate themselves from other, related professions.

In my interview with Steve (the other course instructor), the issue of writing and skills also surfaced. In Extract 3.2, Steve talks about his experience in hiring junior UX writers and the particular (writing) skills candidates should have. Again, the value of this is established through a contrast with the kind of writing other people might do. What I am especially interested in here, though, is the emphasis Steve places not just on writing but also on metalinguistic reflection about one's own writing (cf. Chapter 2).

**Extract 3.2: Writing-as-designing in UX writing course (Steve)**

1. There are a lot of writers out there who, maybe, typically they're, like, journalists,
2. so they're writing articles, and I asked them, "Oh, this is really interesting, why did
3. you structure it this way, and why did you use this language, and who's your
4. audience?" And sometimes people just answer, like, "I don't know, this is just how
5. I wrote it." . . . I don't think if you are a UX writer or a content designer, that's an
6. approach that is acceptable. You're kind of constantly called on A) to kind of
7. rationalize your work and why you made decisions, but B) like, to be able to
8. consistently reproduce those decisions. You have to sort of be consciously
9. thinking about them. . . . A lot of people could be a UX writer kind of implicitly . . .
10. they're good at it, but they don't really– they're not really thinking about the why.

Like Tom (Extract 3.1), Steve uses the example of another writing practice (the writing practice of journalists, in this case), to illustrate how UX writers approach (or should approach) their work. Through this, UX writing is again established as a different – and possibly distinctive – kind of writing, one that is judged against norms different from journalistic writing. This time, however, the main differentiator between the two writing practices lies in the supposedly different metalinguistic reflection about one's work. UX writers, Steve suggests, have to 'rationalize' their work (line 7), which entails both explaining 'why you made decisions' (line 7) and being able to 'consistently reproduce those decisions' (line 8). These decisions pertain to various dimensions of writing and include the structure of the text, word, or register choices, as well as an awareness of audience design (lines 3–4). Evidently, and as Steve himself suggests, other people's writing also entails these considerations; certainly, all writers have to make register and word choices when composing a text, and as Bell's (1984) work on audience design shows, people always orient to a particular audience (whether implicitly or explicitly). Yet, according to Steve, other writers might do this work only 'kind of implicitly' (line 9). What becomes particularly evident here is that this is not a factual description but rather a discursive construction of skills. It seems quite unlikely that journalists do not consciously think about structure, lexis, and audience design; yet, by setting up journalists as 'other' language workers, Steve can make a particular boundary and status claim, thereby positioning UX writing in a favourable light. It is through the contrast with other language workers that Steve can claim metalinguistic reflection as a distinctive skill that distinguishes the kind of writing UX writers do and that demarcates it as a unique professional field.

More broadly, Steve's account suggests that metalinguistic reflection is not only something differentiating UX writers from other writers, but that it is also vital for their status claims vis-à-vis their colleagues (again, recall Chapter 2). In other words, the construction of skills is also used to establish one's position in local hierarchies. UX writers, he explains, are asked to 'rationalize' their work (line 7) and show that they can 'reproduce' previous work (line 8). They are expected to not only reflect on their writing but to also demonstrate their metalinguistic skill by verbalizing and explaining their writing decisions to others. It is a skill UX writers also have to explicitly perform as part of their work (and, as Steve's discussion of hiring practices suggests, it may also be something people have to perform in order to get work in the first place).

Overall, then, writing-as-designing is more than just a practice UX writers engage in; it becomes a socially constructed skill (cf. Steiger, 1993) that can be judged, evaluated, and objectified (cf. Urciuoli, 2008). Importantly, though, it is a skill that is constructed as unique to UX writing, and thus something UX writers can use to distinguish their work and demarcate their professional practice – both internally (vis-à-vis their colleagues) and with regards to other forms of language work. In other words, writing-as-designing is central to constructing the professional field of UX writing. This becomes particularly apparent when one examines how this skill is codified and standardized through textual artefacts, which I turn to discussing next.

### 3.4.2 *Books as Durable Artefacts: Establishing the Professional Field*

Having outlined the central role of writing-as-designing in the professionalization of UX writing, I now turn to examining how this skill is codified or articulated (cf. Håkanson, 2007) in material artefacts such as books about UX writing and how these artefacts in turn also contribute to the professionalization of UX writing more generally. As Smith (2005) suggests, written texts take on a particular role in the standardization of professions. As a (more) 'material thing', written texts have 'constancy . . . that provides for standardization' (Smith, 2005, p. 166); following Håkanson (2007), this can be understood as one way in which knowledge is articulated (i.e. made explicit) and subsequently fixed, standardized, and made replicable. Looking to research in sociocultural linguistics, this aligns with research such as that by Myers (1992) on the social construction of academic texts, who suggests that textbooks, specifically, can be understood as codifying the knowledge of a particular field.

Given the relatively recent emergence of UX writing as a professional field (see Chapter 2), it is not surprising that there are no textbooks or official curricula on the subject at the time of writing. Still, while there is as of yet little institutional regulation on what UX writing is, it is surely no coincidence that multiple trade books on UX writing have been published in the late 2010s. Of these, the books that surfaced as most prominent in my own research were, as mentioned earlier: *Strategic Writing for UX, Writing Is Designing*, and *Writing for Designers*. Affiliated with established publishing houses in software design and engineering, these books in and of themselves constitute a rhetoric of professionalization. *Strategic Writing for UX* is published by O'Reilly Media, known for its extensive range of publications on web and software engineering (including books on design and UX); *Writing Is Designing* is published by Rosenfeld Media, a smaller publisher specialized in publishing books on UX and UX design; and, finally, *Writing for Designers* is published by A Book Apart, an Indie publisher that, as per its own website, publishes 'books that are . . . [a]bout professional work on the web and in the tech industry'.[4] Each of these publishers is a household name in the UX industry. Just as academic journals serve to establish and demarcate new knowledge terrains, these books serve to legitimate UX writing as a professional specialization within the broader field of UX.

Equally interesting is the fact that many of my interviewees explained that the three books on UX writing listed above – as well as earlier books on

[4] Source: *A Book Apart. Pitch us a book.* Retrieved from http://web.archive.org/web/20240221113137/https://abookapart.com/pages/pitch-us-a-book/ (accessed 21 March 2022).

content strategy[5] – had been central in their own education in UX writing. In this sense, I am interested in these books also as part of a broader process that leads to the codification of particular knowledge and skills. These books are a literal materialization of UX writing and its professionalization; it is thus not surprising that the concept of writing-as-designing also pervades them. To illustrate this, I offer Extract 3.3, which shows an excerpt from Podmajersky's *Strategic Writing for UX* where she contrasts UX writing with other kinds of writing.

**Extract 3.3: Writing-as-designing in *Strategic Writing for UX* (Podmajersky, 2019, pp. 137–138)**

1. When most people think of writing, they think of writing words that people will
2. read in a book, article, essay, student paper, or blog. . . . In contrast, UX writing is
3. not a sequence of words, sentences, and paragraphs that stand on their own.
4. Instead, it exists to be the conversation between the experience and the person
5. using it. The experience talks to the person with words and visuals, and the person
6. responds by interacting with elements on the screen. . . . If I were to write the
7. UX text in a written-page format, even in tables or a spreadsheet, I would fail to
8. design for the context.

Like Steve and Tom in the UX writing course (see above), Podmajersky emphasizes that UX writing is different from other typical forms of writing, such as that produced for books, essays, or academic writing (line 2). This is not about prose, indeed; it is not even about words 'on their own' (line 3); rather, it is about highly contextualized writing that constitutes a 'conversation between the experience and the person using it [the software/experience]' (lines 4–5). This kind of writing entails not just words but also visual elements (line 5); owing to this, it – supposedly – cannot be written like any other text but has to be 'design[ed] for the context' (line 8).

Again, the use of 'design' here is particularly telling in a moment where writing is established not in terms of its verbal dimension (i.e. 'words, sentences, and paragraphs', line 3) but through its connection to context (line 8) and the potentially ensuing action by users who 'respond[]' to the software (line 6). Once more, the focus appears to be not on the verbal dimension of

[5] Content strategy is a professional field in web and software design that is generally seen as adjacent to UX writing. The term has been around longer than UX writing and has received particular attention in the late 2000s and early 2010s (for an overview, see Clark, 2016). Content strategy is said to entail more strategic work such as planning and management of content (including long-form content), whereas UX writing is said to be about shorter texts that are part of the user interface. However, these boundaries are relatively fluid; for instance, until October 2020, UX writers at Facebook (now Meta) held the job title 'content strategist'. Source: *Why We're Moving From Content Strategy to Content Design.* Retrieved from https://medium.com/designatmeta/why-were-moving-from-content-strategy-to-content-design-e288a70169b8 (accessed 21 March 2022).

writing but on how writing supposedly should be used to achieve specific ends. In this regard, UX writing is here presented not just as a more-than-linguistic approach but also as a more-than-representational form of writing (cf. Lorimer, 2005; Thrift, 2007). The kind of writing that UX writers do, according to Podmajersky, is not just multimodal (the 'words and visuals', line 5) but goes further by emphasizing users' embodied experience, which results in their 'interacting with elements on the screen' (line 6).

Interestingly, such an understanding of text as something that is experienced by people in an embodied way parallels recent theorization in sociocultural linguistics, where scholars have argued for a more embodied understanding of texts and (renewed) attention to more-than-representational or posthumanist aspects of (interface) texts (e.g. Pennycook 2018a; Jones, 2022b). This, then, is how UX writing is not just about words and text but rather about the actions that occur through the complex agentive chain that connects words, interface designs, UX writers, users, and other agents. I will discuss this more-than-representational approach to writing and software interfaces in more depth in later chapters, and especially in the conclusion to this book. For now, suffice it to say that for UX writers, this posthumanist perspective seems to be linked to a (re)framing of writing as designing.

Still, the use of 'design' in Extract 3.3 is relatively vague. Indeed, throughout the book, Podmajersky does not define what exactly 'design' means and whether 'design[ing] for context' (line 8) is really different from, say, writing for context. As such, Podmajersky's use of the word 'design' (here and elsewhere) might thus also be a way of aligning herself (and other UX writers) with the typically higher-paid field of UX design (cf. Chapter 2). In other words, claiming that UX writers engage in 'design' can raise the status and prestige of the profession.

Without going into much detail, I offer two additional examples of how books on UX writing orient to design as a professional field: Extract 3.4 from *Writing Is Designing* by Metts and Welfle and Extract 3.5 from *Writing for Designers* by Kubie. In both cases, the orientation towards design is of course already made evident in the title of the book; this is further elaborated on in the texts themselves. Extracts 3.4 and 3.5 are both taken from the introduction of the two texts, and they illustrate how writing-as-designing surfaces in each of the books.

**Extract 3.4: Writing-as-designing in *Writing Is Designing* (Metts & Welfle, 2020, p. v)**

In many product teams, the words are an afterthought, and come after the "design," or the visual and experiential system. It shouldn't be like that: the writer should be creating words as the rest of the experience is developed. . . . Writing *is* part of the design process, and writers are designers.

**Extract 3.5: Writing-as-designing in *Writing for Designers* (Kubie, 2018, p. 1)**

"Shit. The writing. We forgot about the writing. The thing, the design thing – it needs words!" ... [W]riting what appears in an interface forces us to name components, articulate choices, and explain things to the user. It's part of the design.

Similar to how Podmajersky (Extract 3.3) emphasizes that UX writing is different from other 'traditional' forms of writing, these two extracts also point out that writing in this instance is not just about words, but that it entails a particular focus on context. It 'should be creat[ed] ... as the rest of the experience is developed' (Extract 3.4, line 3), and it attends not just to verbal aspects but also centrally includes 'articulat[ing] choices, and explain[ing] things to the user' (Extract 3.5, line 3). In short, UX writing is not just about writing, it is 'part of the design' (Extract 3.4, line 4; Extract 3.5, lines 3–4).

What is particularly evident in these two extracts is again how the word 'design' can be used to affiliate oneself with this adjacent professional role. In both extracts, the authors lament how words are often forgotten in software design: Metts and Welfle (Extract 3.4) explain that software design teams often see words as an 'afterthought' (line 1); meanwhile, Kubie (Extract 3.5) constructs a scenario in which an imagined team ('we', line 1) completely forgot that 'the design thing ... needs words' (lines 1–2). In light of this, claiming that writing is part of the design might be a strategic way of positioning oneself as part of a particular 'process' (Extract 3.4, line 4) in order to gain more prestige and more institutional impact. I will return to this aspect later when discussing job titles; first, however, I want to consider how books like these also contribute to the codification or standardization of specific skills as typical for or inherent to UX writing and how this in turn contributes to establishing the professional field of UX writing.

As artefacts with relatively durable materiality, books can serve both as a record of industry knowledge and as an index to this expert knowledge. Indeed, the consolidation of knowledge can be what leads people to create such artefacts in the first place, although such acts always also entail aspects of profession building and of demarcating one's own field. To exemplify this, I turn to the preface of Podmajersky's *Strategic Writing for UX*, an excerpt of which is reproduced in Extract 3.6.

**Extract 3.6: Profession building in *Strategic Writing for UX* (Podmajersky, 2019, p. xi–xii)**

We UX writers haven't had a common set of frameworks, tools, or methods that address the unique challenges of UX content. The organizations and managers who want to hire us might know they have a "word problem," but they have a hard time figuring out who to hire, how to support us, and what impact to expect. This book was conceived when I realized that we can't have a community or discipline of UX writing until we hold some basic ideas in common. We need to share expectations for what UX content can do, best practices for making the content do what it can, and methods to measure its effect. I wrote this book to share my frameworks, tools, and methods for creating UX content, and to share my encouragement and enthusiasm for using UX content . . . .

Here, Podmajersky reveals her profession-building intentions by demarcating the problem she sees, namely that UX writers do not (or not yet) have 'a common set of frameworks, tools, or methods that address the unique challenges of UX content' (lines 1–2). In other words, there is no shared knowledge system uniting UX writers as a professional group (cf. Abbott, 1988). As she suggests, writing this book was a way for her to share her 'frameworks, tools, and methods' (line 8); Podmajersky's book thus becomes a durable record of her own knowledge and expertise in UX writing. More interesting than this, however, is her profession-building agenda. The book aims to solidify the professional knowledge of UX writers by making the tacit knowledge that an individual practitioner has (their 'knowing-in-practice', cf. Schön, 1993) explicit. As a well-read book within the UX writing community, *Strategic Writing for UX* ultimately contributes to establishing what counts as shared knowledge, which in turn is key for the production of a professional field.

Interestingly, Podmajersky places also significant emphasis on community in this preface. Not only does she start with 'we UX writers' (line 1) and repeatedly use the pronoun 'we' throughout the text, but she also creates a clear distinction between the in-group of UX writers, and organizations and managers as an external other (lines 2–4). This outgroup is characterized as unfamiliar with the work UX writers do, aware only of a 'word problem' (line 3); note how the use of the scare quotes underscores the inability of the outgroup to understand the actual work UX writers do and the skills that are necessary for this work. At the same time, the sense of an ingroup is strengthened through lexical items emphasizing the notion of 'community' and 'sharing'. The text makes explicit reference to 'community' in line 5; it also stresses the idea of 'common' (i.e. mutually shared) knowledge at various points: UX writers are said to be in need of a '*common* set of frameworks, tools, and methods' (line 1; emphasis added), and they ought to 'hold some basic ideas in *common*' (line 6; emphasis added). Equally, 'sharing' is repeated several times: Podmajersky urges people to 'share expectations' about what UX writing can do and how one is supposed to do it (lines 6–7), and she declares that her own reason for writing this book was 'to *share* [her] frameworks . . . and *share* [her] enthusiasm' (lines 8–9; emphasis added). What emerges is a particularly strong sense of community and community building.

I also encountered a strong sense of community at various other moments in my research on UX writing; this parallels the importance Kou and Gray (2018) attribute to communities of practice in their discussion of the professionalization of UX design. Similarly, Kong (2014) sees professions as 'symbolic communities', which are defined by 'shared membership, knowledge, ideology, values, interpersonal positions, activities and resources' (p. 49). Importantly, Kong suggests that such symbolic communities are constructed and bound together through the use of symbolic resources like language. Thus, the professional field depends also on a sense of shared dispositions, which might in part be mediated through artefacts like books.

Texts are, however, not just fixed entities; their use and meaning also depend on who is reading or interpreting them. Orienting to Smith's (2005) discussion of the ability of written texts to coordinate actions and social relations in professional contexts, Kong (2014) therefore sees a written text as something that 'can be stored as a permanent record for a long time but is subject to individual use and interpretation' (p. 5). In this sense, codified professional knowledge is never fixed but depends on the social context in which it is invoked (cf. Kong, 2014), and one could expect that the knowledge and skills codified through books on UX writing are interpreted and applied differently from practitioner to practitioner. What interests me most, however, is how these books themselves can become socially meaningful as symbolic objects that practitioners can use to engage in boundary work. In this sense, books can be used to associate oneself with a particular professional practice (cf. Porcello, 2004). To illustrate this, I offer Extract 3.7 from my interview with Steve, one of the two instructors of the UX writing course. The exchange in Extract 3.7 occurred after I had asked Steve about his own professional trajectory. I am interested in this interaction partly because of Steve's own story and the emphasis he himself places on books in the interaction that ensues; what I find especially intriguing, however, is the moment of recognition that occurs between Steve and me as he mentions these particular books.

**Extract 3.7: Profession building in interview data (Steve)**

| | | |
|---|---|---|
| 1. | **Steve** | kind of early on I read uhm *The Elements of Content Strategy* by Kerin- uh |
| 2. | | ((shakes head)) by Erin [Kissane (.) ] |
| 3. | **Lara** | [((nodding vigorously and smiling))] [mhm] |
| 4. | **Steve** | [uhm ] |
| 5. | | which is e- extremely foundational like there's so: many [people ] |
| 6. | **Lara** | [((nodding))] |
| 7. | **Steve** | who have been doing this for about (.) ten-ish years like I have (.), who uhm |
| 8. | | (.) say that that book just like taught them so much=>between that and |
| 9. | | Kristina Halverson's book< [it's just like was] |
| 10. | **Lara** | [((nods once)) ] |
| 11. | **Steve** | (.) really useful and influential |

As mentioned, what interests me most about Extract 3.7 is the moment of recognition that occurs in this interaction. This starts with my response to Steve's mention of *The Elements of Content Strategy*, which consists of vigorous nodding followed by an additional verbal response token ('mhm', line 3). At first sight, this stretch of nodding may not seem particularly noteworthy; nodding is a common backchanneling strategy (e.g. Schegloff, 1982) which might be particularly important in video-mediated interaction, where participants' visual and embodied participation can take on a more central role (cf. Due & Licoppe, 2020). The turn on line 3, however, constitutes relatively prolonged nodding (four complete nods, to be precise) and becomes a marked acknowledgement of Steve's previous turn. Additionally, the placement of this response is telling; it follows Steve's self-repair after mispronouncing the author's first name and overlaps with his pronunciation of the author's last name. This overlap likely signals an understanding of the trajectory of Steve's turn (cf. Liddicoat, 2011, p. 117), and, in this way, my own knowledge of the book in question.

The shared recognition of UX writing knowledge becomes even more evident as the conversation continues. On lines 8–9, Steve mentions a second book about content strategy; this time, however, he does not mention the actual title of the book but refers to it simply as 'Kristina Halverson's book'.[6] Through this, he presupposes the title of the book as common knowledge for me. Steve is also noticeably speeding up his talk as he mentions this second book, possibly signalling that this information is not new but already known – by me and/or people in the UX writing community. Again, my non-verbal response (line 10) follows immediately after Steve mentions this book, this time less marked, indicating perhaps that common ground has already been established at this point. (I did, in fact, know both books he was referring to.) The recognition in Extract 3.7 thus constitutes a recognition in two senses: on the one hand, my nodding signals prior knowledge of these two books; on the other hand, this interaction also constitutes a recognition in the sense that our alignment on the matter produces the books in question as part of a canon that UX writing professionals (should) have read. This recognition of a UX writing canon establishes once again an in-group, signalling that there is a professional community of UX writers who hold particular understandings of what UX writing entails, including which books codify and standardize that knowledge.

More broadly, the repetition of known books about UX writing Steve and I engage in here also creates a narrative of coherence, establishing UX writing as a phenomenon with a consistent history. This is not unique to the interaction in Extract 3.7. As mentioned before, several of my interviewees talked about

[6] The book in question is *Content Strategy for the Web* by Kristina Halverson.

books like Podmajersky's *Strategic Writing for UX* in interviews (though they sometimes assumed I did not know these books), and these books were also frequently featured in social media posts I encountered. As such, trade books may be used to signal affiliation with UX writing as a professional community and to establish an overall narrative of the profession; it is through the frequent repetition of such books that the field becomes more stable and recognizable to in- and outsiders.

Written texts like books thus serve several important functions in the professionalization of UX writing. They can be understood as material artefacts that form part of the abstract knowledge system of UX writing as a professional group (cf. Abbott, 1988), constituting an explicit articulation and codification of professional knowledge and skills: it is here, for instance, that *writing* in UX writing is explicitly anchored as writing-as-designing and formulated as part of the skillset that is attributed to UX writing. As such, these books always also create particular agential cuts (cf. Barad, 2007) through which the professional identity of UX writers is produced. Importantly, then, these books are never just fixed entities; they are deliberately used by UX writers as indexical means for creating their professional identity in situ, for instance by signalling affiliation with the professional field of UX writing (cf. Extract 3.7). More broadly, these books also serve a profession-building agenda: by establishing what counts as shared knowledge within UX writing, these books can demarcate the boundaries of the professional field. Writing, reading, or mentioning these books is thus enmeshed with particular professional roles and identities that are tightly connected to claims about the profession itself and to legitimating UX writing as a professional specialization within the broader field of UX.

### *3.4.3 Job Titles: Naming and Negotiating Professional Affiliation*

UX writers' efforts to establish themselves as equal to other professions in UX are also especially apparent in the job titles these people chose for themselves. Before, I discussed how books may be used to signal professional affiliation; such indexing of professional affiliation is, of course, even more evident in job titles. Once again, UX writers' orientation to writing-as-designing is emphasized here.

What is striking about the professional field of UX writing is that there is considerable debate around whether 'UX writer' is indeed the right title for the profession. Throughout my research, I regularly came across such discussions: in industry newsletters, on social media sites, or in casual conversations with people who worked in UX and knew about my research. (It was not uncommon for people to ask me about what, in my professional opinion, the 'best' title for this kind of work was.) Around 2021, I noted a veritable surge in blog posts published by UX writers affiliated with well-known software companies such as Meta or Uber, who explained why and how their team had changed its name to a new one (e.g. from 'UX writers' to 'content designers'). To be clear, these various professional titles had been around for a while at that time, and as discussed in Chapter 2, in my interviews with UX writers, many of them explained that their own title had not always been UX writer (or that it in fact wasn't UX writer at the time of the interview). While some of my interviewees had strong opinions on the topic, others consider the search for the perfect professional title, and especially the fine-grained, sometimes arbitrary (or contradictory) distinctions people made between different labels pointless. Regardless of this, however, job titles do of course have implications for those that hold them.

While there is little work on the specific role of job titles in sociocultural linguistics, it is clear that the metapragmatics of naming more generally (or onomastics; see, e.g., Puzey & Kostanski, 2016) play a significant role in the construction of social reality and that the question of who gets to name what is not just a question of labelling but also a question of politics and power (cf. Karlander, 2017, who discusses this with regards to place names). Inevitably, the naming of one's work is tied to practical, institutional concerns, as well as to identity performances. As scholars in organizational studies point out, job titles can be seen as 'identity badges' that define people in professional as well as private contexts (Grant et al., 2014; see also Baron & Bielby, 1986). Outside of work, they are – at least in a Western context – an important vehicle for self-presentation; at work, they are often seen as (supposedly) encapsulating the skills and expertise an employee has.

Taking into account the social significance of job titles, I am intrigued by how job titles in UX writing are indexicalized on various levels (Silverstein, 2003; see also Landqvist, 2019) and how such indexical meanings are deployed. 'UX writer', for instance, points to the field of UX (as discussed in Chapter 2) and thus allows people to situate themselves within this larger tradition. But the indexical field may also be smaller, pointing instead to a particular team or organizational unit an employee is situated in (for instance, the UX team rather than the marketing team). As Landqvist (2019) suggests in his writing on indexicality and names, there is never just one indexical context for a term; rather, contexts are parallel, 'occur[ing] in different times and places' and sometimes in a contradictory manner (p. 727). In this sense, names and naming can be understood as indexical practices with very concrete, interactional aims and at the same time also as something that can function as symbolic or linguistic capital (cf. Tan, 2021); in other words, names can have immediate or local as well as more long-term, strategic purchase.

While people may have rather strong opinions on which job title they prefer to use in a professional context, most of my interviewees were perfectly content with calling themselves simply 'writer' when talking to people outside their industry. However, whenever I explicitly asked about job titles, they seemed to be quite aware of the politics of such labels and their indexical power. Consider Extract 3.8, where Lisa (pseudonym), one of my interviewees, explains why she chose the title 'UX writer' when joining a new company. Lisa has been working as a UX writer for over a decade and considers herself 'maybe one of the people who have done this the longest in the industry'.

**Extract 3.8: Job titles in interview data (Lisa)**

We landed on UX writing because we thought it was better to have the word "writing" in there. The alternatives like "content strategy" and "content design", content strategy, no one knows what that is. It was awful when I was called a content strategist, no one knew what it was. . . . It sounded too high level. Like, "strategy" gives that connotation, but the actual work of it was not seen as important, so there's a tension there. . . . I would rather be called a UX writer than a content strategist, because it just makes my actual day-to-day life easier if people see me as a writer rather than as a strategist.

Lisa's account suggests that her issue with the job title content strategist lies in what 'strategy' indexes: 'it sounded too high level' (line 4). She identifies a tension between this and her actual work, which 'was not seen as important' (lines 5–6). As such, there seems to be a mismatch between what 'strategy' points to and what Lisa actually did (or was allowed to do) as part of her work; as a result, the job title became too obscure, so that 'no one knew what it was' (line 4). By the same token, Lisa's reasoning for choosing UX writer as her new title is that 'it was better to have the word "writing" in there' (lines 1–2), and thus to have an indexical link to the broader field of 'writing' as a professional activity. Lisa indicates that this is more easily understood by colleagues; as she says, 'it makes my actual day-to-day life easier if people see me as a writer' (lines 7–8).

As discussed above, 'writing' is perhaps not such a straightforward term as one might think (see Lillis, 2013); nonetheless, my interviews with Lisa and other UX writers suggested that there seems to be a general understanding that a writer is someone who concerns themselves with words and sentences. As another interviewee explained, when hearing the title UX writer, 'you think about the person who writes'. As a result, a job title like UX writer can be more easily deployed for practical understanding in everyday collaboration with others, as Lisa's example shows. In this sense, there is a direct or first-order indexicality (Silverstein, 2003) to job titles, whereby a job title describes a person that does a particular kind of work. Choosing the label 'writer' can therefore be a way for UX writers to signal the day-to-day work they do to their colleagues.

Next to practical considerations, job titles also encapsulate how people position themselves: how someone relates to other actors and activities, and how they are in turn perceived by other people. Job titles therefore also have higher order indexicality, whereby they index particular professional identities and practices (again, see Silverstein, 2003, on indexical order). It is in this way that job titles become a matter of politics and status. Indeed, while Lisa chose the label UX writer because it indexes writing, others I interviewed considered exactly this aspect problematic, finding that it reduced them to being perceived as 'the words person'.

In the end, any label categorizes people in a way that indexes some aspects of that group but not others (Duchêne, 2020). What became clear in my interviews, though, was that people chose specific labels as much (if not more so) for how they were perceived by others than how they saw their own professional identity reflected in a specific title. Several of my interviewees thus emphasized their preference for 'content designer', which Lisa mentions but does not comment on. My other interviewees explained that using the title content designer mattered to them because it had the word 'design' in it, which would allow them to be seen as part of a company's design team; this in turn could give them access and influence they otherwise would not have.

The symbolic investment in the label 'design' also became apparent in my interview with one of the instructors of the UX writing course, Steve, where the topic of professional labels arose spontaneously. Extract 3.9 shows his thinking on the politics behind *not* choosing the title UX writer.

**Extract 3.9: Job titles in interview data (Steve)**

1. A lot of times at tech companies, writers are undervalued and so if you're
2. assigning "writer" as part of your identity . . . you're sort of internalizing that
3. corporate undervaluing of you and strategically, that's one of the reasons I didn't
4. want to, like, have our team be UX writers, because I wanted us to sort of be in the
5. same pay band as designers.

Other than Lisa (Extract 3.8), who chose the title UX writer for her team, Steve explains that he deliberately chose *not* to use this label. Unlike in Lisa's case, though, his reasons are not linked to the direct activity the term indexes but rather to how the status of 'writer' as a professional category tends to be perceived: writers tend to be undervalued at tech companies, he explains (line 1). This is less about establishing one's professional field vis-à-vis other forms of language work and more so about establishing it vis-à-vis other roles that are internal to the field of UX. Job titles in this case are not so much a matter of practicality and more so a matter of symbolic and, ultimately, material capital (cf. Bourdieu, 1986) – in other words, job titles might have a direct impact on one's salary.

Steve's account here also illustrates nicely how such different concerns around job titles are never neatly separated. Notice how he starts by connecting the question of job titles first to identity and to how people 'internaliz[e]' an undervalued identity (lines 1–3) before linking job titles more explicitly to issues of material compensation by mentioning pay bands (line 5). Identity and institutional concerns appear to overlap. Therefore, the preference for ‘designer’ that many of my interviewees expressed may partially have to do with the cachet of certain kinds of design (cf. Walker, 2009), but as Steve’s example illustrates, the issue is always also linked to more concrete and often material concerns: it is not just that people want to be perceived as designers for reasons of prestige, but also (or perhaps primarily) because this might directly lead to economic capital. For my interviewees, calling themselves ‘designer’ was a way to strategically affiliate themselves with a professional field that came with more influence and higher financial compensation.

To summarize, job titles thus really are a form of linguistic capital (cf. Tan, 2021), one which may translate into both symbolic *and* material capital. They are a way for UX writers to index particular skills, which, in turn, enables them to negotiate the status of their work both with regards to practical, interactional aims and in terms of their symbolic status and financial compensation. Additionally, these often-public discussions of job titles might also be understood as a status-producing mechanism more generally. In the end, being able to choose one’s name is, of course, always a matter of power and privilege, and public debates about a profession’s label always also mark something as a profession in and of itself, as a field that can and, it would seem, must be described in the first place.

## 3.5 Conclusion: Skilling the (UX) Writer

> The notion of profession is dangerous because it has all appearances of false neutrality in its favor. Profession is a folk concept which has been uncritically smuggled into scientific language and which imports in it a whole social unconscious. (Bourdieu & Wacquant, 1989, pp. 37–38)

I started this chapter with two assumptions: first, that UX writing is a profession and, second, that it is a relatively prestigious profession. In this chapter, I have sought to nuance these assumptions by exploring how UX writing is constituted as language work and, specifically, how this kind of language work is legitimized and professionalized. As Bourdieu (quoted above) reminds us, profession is not a neutral label; far from it, it is a concept whose ‘false neutrality’ may fool academics and non-academics alike. Writing around the same time, Abbott (1988) succinctly summarizes the problem: ‘[p]eople don’t want to call automobile repair a profession because they don’t want to accord it that dignity’ (p. 8). In other words, profession is a category that is somewhat arbitrarily bestowed upon some groups and not others. This bears two significant implications, namely that professions are discursively constructed and that the concept of profession is closely tied to status and prestige. In this chapter,

I have sought to address both of these issues vis-à-vis the language work of UX writers, exploring how UX writers discursively professionalize their own work. My focus has been on skill discourse (cf. Urciuoli, 2008) and on the construction, codification, and indexing of writing-as-designing as a unique skill that allows UX writers to establish the (more) privileged status of their work.

Concretely, I have discussed three distinct moments or rhetorics of professionalization. First, I have suggested that the discursive construction of 'writing-as-designing' as a unique skill is a way for UX writers to negotiate the boundaries of their work. This skill entails an understanding of writing as a technology or cultural tool (cf. Lillis, 2013); as a rhetoric of professionalization, it also depends on UX writers' ability to metalinguistically explain their own writing. It is through this metalinguistic work that writing-as-designing becomes a socially constructed skill (cf. Steiger, 1993) that distinguishes UX writing as a unique professional field. Second, I have suggested that skills like this become codified in written texts such as books on UX writing. Books on UX writing therefore function as a material reification of the abstract knowledge system that constitutes UX writing as a professional group (cf. Abbott, 1988); at the same time, books themselves are also cultural tools that can be deployed to signal professional affiliation and to establish a narrative of coherence about the professional field itself, thereby serving a profession building agenda. Finally, I have shown how job titles are also implicated in discourses of professionalization. Through job titles, skills can be indexed; as such, job titles can be a way to express affiliation with specific activities and with particular prestigious and high-paying professional domains. All of this illustrates nicely how the emergence of the professional role and identity of 'the UX writer' is subject to complex configurations of agential cuts (cf. Barad, 2007) and thus not pre-existing to any discourse but defined through material-discursive enactments of the sort described in this chapter. More specifically, the three material-discursive rhetorics of professionalization described in the chapter are interconnected through their focus on skill and, specifically, through their conception of writing-as-designing as a (unique) skill. Ultimately, this skill discourse constitutes a form of symbolic capital that can be deployed to establish the privileged status of UX writing as elite language work.

While research on language work has engaged with the issue of skills and skill discourse (e.g. Urciuoli, 2008; Allan, 2013), the topic of professionalization is discussed less often. This is likely tied to the field's (past) orientation to lower-status language work (cf. Chapter 1); still, what I hope to have shown in this chapter is that a particular focus on skilling and professionalization might enable scholars to better understand not just the language work of UX writers but also what becomes legitimized as (language) work more generally. In their substantive review about research on language and/as labour, Urciuoli and LaDousa (2013) suggest that whether language 'is' work ultimately depends not on the form that it takes but on its function, on 'whether linguistic practices are . . . performed as labor' (p. 176). Yet, evidently, not all language work is equal. The work of a call centre worker clearly differs from that of a UX writer, not only in terms of the specific linguistic practices that are involved but especially in terms of how this language work is valued. As Duchêne (2020) summarizes, while more elite language workers and what he calls 'parole-d'oeuvre' (i.e. low-status language workers) have in common that language is central to their work, their social position and status differ. Language work is therefore entangled not just with people's beliefs about *language* but also – or especially – with their beliefs about what counts as (professional) *work*. In this regard, and as I hope to have shown in this chapter, focusing on skills and professionalization can be particularly insightful in order to understand different hierarchies of language work.

In looking at professionalization, (language) skills can thus be understood as a form of linguistic capital (Bourdieu, 1977); relatedly, the discursive construction of skills can be understood as an 'appeal to professionalism' (Fournier, 1999) that elite language workers like UX writers deploy in order to claim status and expertise in both local and wider hierarchies of language work. As my case study illustrates, these local and wider hierarchies of (language) work are connected to the establishment of a professional field (cf. Bourdieu & Wacquant, 1989) and to people's positioning vis-à-vis other fields; it is in this way that an at times explicit hierarchy is established, whereby some kinds of language work are valued more than others.

Whether judgements about the value of language work are seen as justified or not is, of course, a matter of debate and has less to do with the concrete linguistic work that is done and more so with particular status claims (cf. Duchêne, 2020). Thus, these hierarchies between language work are established not through how linguistic products or even language itself is valued, but rather through how specific ways of thinking about, understanding, and using language are constructed. In other words, it is skills and skill discourses that are deployed to demarcate difference; this makes skilling and professionalization particularly helpful concepts for understanding how high-end language work differs – or is constructed as different – from other kinds of language work. As such, the current chapter has shown that the status of UX writers as *elite* language workers depends on their relative positioning and on differentiating themselves from other language workers. I expand on this topic in the chapter that follows, where I examine UX writers' status claims with a particular view to how they (dis)avow creativity as a professional value.

# 4 Claiming (Non-)Creativity

## The Symbolic Economy of Creativity in UX Writing

> To say something is "creative", is, above all, to make a value judgement, and the value we assign to something by calling it "creative" is not just a judgement about its formal qualities or the relative skill it took to make it. It's a judgement about its social worth, about its place within a particular moral universe. (Jones, 2018, p. 82)

In Chapter 3, I discussed how UX writers demarcate their own profession by differentiating it from other kinds of writing and other kinds of language work. In this regard, UX writing is, as I have shown, often described as being concerned more with the effect or function of words rather than their form. Indeed, some UX writers even explicitly state that UX writing is not about 'creative expression' (cf. Extract 3.1 in Chapter 3). Why then dedicate an entire chapter to UX writing and creativity? The first reason has to do with the broader professional categorization of UX writers, who (along with other professionals in software design, such as designers or engineers) are typically labelled 'creatives'; they sit at the core of Florida's (2012) infamous 'creative class'. Second, and more importantly, claiming that something is *not* creative is not the same as an absence of creativity – indeed, my point is that invoking non-creativity always also entails positioning oneself with regards to other, usually positively connoted discourses of creativity. In other words, such claims are always connected to particular value judgements (see Jones, quoted above). This is the issue that sits at the heart of the current chapter, where I discuss the different ways in which UX writers invoke creativity and non-creativity with regards to their work.

By examining some of the claims UX writers make about their own work as being creative or not, I reflect on how creative language depends as much on semiotic ideologies (Keane, 2003, 2018) as it does on linguistic properties. Indeed, following Thurlow (2019), I argue that creativity is not just a technical, linguistic accomplishment but also a discursive strategy or rhetoric. In this way, I consider how UX writers' claims to (non-)creativity allow them to construct their identity as competent language professionals and negotiate their position in the relative hierarchies of language work more generally. Through this case study, I seek to problematize both human-exceptionalist approaches to creativity as well as those that seek to democratize creativity. Contrary to such approaches that try to define the nature of creativity (or to expand existing definitions), my concern is rather that what counts as creative or not – as well as the value that is attached to this label – is subject to particular configurations of material-discursive enactments (or agential cuts, cf. Barad, 2007) that contribute to the emergence of the kinds of subjectivities and professional identities described in this chapter.

The chapter is organized as follows. To start, I give a brief overview of how creativity is or has been commonly understood, before considering how the topic of language and creativity has been discussed by sociocultural linguists more specifically. After this theoretical framing, I turn to discussing how UX writers, as a typical example of contemporary wordsmiths, invoke creativity when talking about their own language work. Drawing on language biography interviews with UX writers as well as a handful of second-hand, public interviews with UX writers, I discuss how these elite language workers strategically deploy discourses of creativity to fashion their professional identity. In my analysis, I show how it is not just their claims to creativity but also – or perhaps especially – their claims to *non*-creativity that matter for demarcating both their language work and their position or status vis-à-vis other language workers. In this sense, I end by linking my case study to the broader question of status and privilege in language work, arguing that an analysis of the rhetorics of creativity can help sociocultural linguists better understand hierarchies within and between different kinds of language work.

## 4.1 Key Concepts: Language, Discourse, and Creativity

Creativity is difficult to define. While most might agree that creativity is something good – as Toolan (2012) remarks, '[w]e would all prefer to be creative rather than not' (p. 19) – what exactly counts as being creative is usually less clear-cut. Particularly (but not only) in the past, creativity has often been conceptualized as artistic or high-cultural output, with creativity itself being seen as 'the solitary, ex nihilo creation of products of self-evident and universal value ... by highly exceptional and gifted individuals' (Wilf, 2014, p. 398). In this view, creative agency is located in single individuals. While such views of creativity are millennia old, they have by no means disappeared: Wilf (2014) suggests that the idea of the Romantic genius lives on and today takes the form of a 'neoliberal sequel', which emphasizes an individual's ability to imagine the new and the unknown – in short, to innovate. Such views of 'exceptional' creativity have rightly been critiqued by many scholars. Wilf himself notes that such views obscure the social dimensions of creativity and ignore the fact that creativity depends as much on communicative context, that is, on how people are socialized into creativity and on how a society recognizes acts of creativity, as it does on the (presumed) creative work of the individual.

Sociocultural linguists, too, have critiqued perspectives that locate creativity in the spectacular and the extraordinary. This 'democratisation of creativity' (Maybin & Swann, 2007) is perhaps best encapsulated in Carter's (2004) proclamation that 'linguistic creativity is not simply a property of exceptional people but an exceptional property of all people' (p. 13). In this vein, many sociocultural linguists have argued that creativity can also be found in the most mundane communicative practices (e.g. Maybin & Swann, 2007; Atkins & Carter, 2011; Swann & Deumert, 2018). Still, not all creativities are equal. As Jones (2012) argues, while one might rightly label a cleverly improvised excuse as 'creative', this kind of creativity is still different – or seen as such – from the kind of creativity one would attribute to the work of an esteemed novelist or composer. Several scholars have tried to explain this difference by making a distinction, for instance, between improvisation and innovation (Liep, 2001), or between personal creativity and historical creativity (Boden, 2004). However, while such categorizations are interesting to consider, they are often too broad and too generalizing for the kinds of analyses sociocultural linguists are typically interested in.

In what follows, I thus briefly outline three different ways in which language scholars can and have looked at language and creativity: (a) as linguistic creativity; (b) as discursive creativity; and, finally, (c) by looking at creativity as a discursive strategy or rhetoric. I do not mean to suggest that these approaches are mutually exclusive. Rather, they each approach creativity at a different level, allowing scholars to answer different kinds of questions about language and creativity. At the same time, my subsequent analysis shows that UX writers, too, understand creativity as something that has various nuances – indeed, their own understanding of creativity is often not so different from how sociocultural linguists have conceptualized creativity.

### *4.1.1 Linguistic Creativity or Creativity as a Product*

Early research on language and creativity often approached creativity through the lens of stylistics and the analysis of literary texts (for an overview, see Jones, 2010). In such research, creativity is presumed *a priori* and taken for granted as a feature of the texts analyzed. This kind of research falls into what Jones (2010) calls the paradigm of 'linguistic creativity', where creativity is sought in the specific form of language and where the focus lies on the creative product (for instance, the literary text). Such an approach, however, is not reserved for the analysis of exceptional works of creativity. Recall Carter (2004), who argues that literariness can be present in any kind of text or talk. Indeed, in his work, Carter traces the occurrence of features such as wordplay, rhyme, puns, repetition, or metaphor in everyday spoken English and demonstrates that literary-like language is present in ordinary talk as well.

More importantly, though, Carter also shows that literariness in everyday interaction can perform social functions and that people use such linguistic features to build and sustain relationships (see Atkins & Carter, 2011). In this sense, research such as that of Carter and his colleagues shows how creativity always occurs in a specific social and communicative context. Still, such research conceptualizes creativity primarily as a linguistic achievement. As Jones (2010) puts it, 'researchers in this paradigm tend to focus on the social functions of *creative language* (by which they usually mean "literary" or "poetic" language), rather than the function of language (of all kinds), in performing *creative actions*' (p. 469, emphasis in original). To put it differently, research in the linguistic creativity paradigm retains an emphasis on creative linguistic products while, for the most part, ignoring creative processes.

### *4.1.2 Discursive Creativity or Creativity as a Process*

In contrast to research on linguistic creativity, Jones (2010, 2016) advances an approach to creativity that puts discourse and discursive action at the centre. As he proposes, one can look at creativity not just as a product but also as a process: creativity can also be about the *use* of language and the creative actions people can take through language. Creative practices are then seen as contingent and emergent, as something that is, as Swann and Deumert (2018) point out, 'responsive to times, places and social relations, but also transformative of these' (p. 6). Needless to say, this demands going beyond the (creative) text and focusing on what happens around it. Creativity in this sense is about how communicative practices occur in particular contexts and interact with these contexts.

A discursive approach to creativity demands taking into account social and historical dimensions, as well as the ways in which texts are (re-) contextualized interdiscursively (see Jones, 2010). Jones (2010) exemplifies this way of 'reading *around* the text' (p. 477, emphasis in original), by way of analyzing a one-line slogan used in a governmental anti-drug campaign in Hong Kong, '*sang meng mo* Take 2' – 'There's no "Take 2" in life'. Through direct involvement of the intended recipients of the campaign, Jones is able to show how the slogan is creatively appropriated by them. More than that, though, he shows how their creative transformation of the slogan into 'There *is* a take 2 in life' is ultimately not only more effective than the government's campaign, but it also opens up a space for the recovering drug users to contest the negative and damaging discourse imposed by the original slogan. For Jones, a discursive approach to language and creativity should thus be interested in 'how language can be used in strategic ways in concrete social contexts to create fundamental changes in the world' (p. 468). Such changes may surface at the level of local interactions between participants, but they may also occur on a higher level, potentially affecting cultural change.

Jones's (2010) example also highlights – or complicates – another aspect of creativity: the often-taken-for-granted assumption that creativity is necessarily located in the realm of joy and pleasure, disconnected, one might think, from issues of politics and power. The question of power, however, surfaces quite clearly when one looks at what kinds of creativities are deemed worthy and/or valuable. Thurlow (2012) shows this nicely in juxtaposing the everyday creativity of young people's language play in digital media with the uptake of the same linguistic forms in advertisements. While young people's creative and playful practices of new media language are ridiculed and demonized by news media, advertisers can recycle the same kind of language play for commercial gains, effectively commodifying young people's creative practices. Thurlow thus shows not only that the cultural politics of creativity are connected to moral values but also that such judgements can translate into economic value – and, not surprisingly, the question of whose creativity is valued and whose is not usually turns out to be as much about the creators as it is about the creative object. Seen this way, creativity comes about through culture and interaction, and it is enmeshed with cultural-political judgments. This leads me to a last and third approach to creativity: how creativity might be thought of as a rhetoric or discursive strategy.

### *4.1.3 The Rhetoric of Creativity or Creativity as a Discursive Strategy*

Declaring that something or someone is creative always involves making a value judgement (Jones, 2018, p. 82). This raises not just the question of how people and things are ascribed creativity, but also why and with what effect creativity is claimed or assigned. This is where Thurlow (2019) sees an additional perspective or angle on language and creativity: the way creativity itself may be deployed as a discursive strategy, as something that is invoked 'in specific moments, by specific agents, and for specific ends' (p. 98). Such a meta-level approach to creativity is less concerned with creative products (i.e. linguistic creativity) or with creative actions (i.e. discursive creativity) and instead attends to judgements of and claims to creativity.

While not directly concerned with creativity as a label, Norris's (2012) detailed account of how a community artist comes to be recognized *as* an artist neatly illustrates how creativity can work as a rhetoric. In her case study, Norris discusses how some actions produce one social actor, Andrea, as a community artist, while the same kinds of actions, when performed by other people, do not have the same effect. What her analysis illustrates is that

Andrea's ability to claim a creative identity is not just determined by what she does but also – and perhaps as much – by the views the people surrounding her have of what it means to be a creative person. In other words, Andrea is not (or at least not only) creative because of what she does, but she is produced *as* creative by being ascribed creativity. Similarly, Wilf (2014) argues that art and artists are created through the discourse surrounding them, noting that both 'can become singular as a result of their carefully entextualized biographies' (p. 405). Simply put, then, creativity itself is something that is discursively accomplished. Building on this, creativity can be understood as a resource people can draw on to fashion their own or someone else's identity. This in turn raises the question of how creativity may be deployed strategically, or, as Thurlow (2019) suggests, it 'obliges us to recognize creativity as a (meta-)cultural and cultural-political (or ideological) phenomenon' (p. 98). What becomes of interest is who claims to be creative, when, how, and, most importantly, why. In other words, the focus shifts from creative use of language/discourse to creativity as a rhetoric.

If one understands creativity as a rhetoric, it becomes impossible to decouple creativity from issues of power and politics. In this way, the readiness of scholars to celebrate everyday creativity is in and of itself also a value judgement (see Thurlow, 2019). By the same token, if conceptualizations of creativity as 'exceptional' live on in the modern 'genius' and in neoliberal notions of innovation (Wilf, 2014), then scholars need to look not just at everyday creativity but also at how and when lay people declare that something is (exceptionally) creative. I do not mean to suggest that linguistic and discursive creativity do not matter – indeed, as I will show in my analysis of creativity in UX writing, they matter quite a lot for people in order to make sense of creativity for themselves. What I am suggesting, however, is that this additional perspective on creativity – of interrogating creativity as a rhetoric in and of itself – can be beneficial for understanding how creativity in its entirety operates and how it structures people's social and material worlds. With this in mind, I now turn to considering the case of creativity in UX writing, discussing in particular how creativity is leveraged as a symbolic resource with potentially material outcomes – and this includes the way it may be denied or disavowed.

## 4.2 The Data

For my case study, I draw primarily on interviews I conducted with 21 UX writers about their own use and understanding of language as part of their work. As outlined in the introduction to this book, I call these interviews *language biography interviews*, since they emphasize individual and affective perspectives that are connected to people's unique professional trajectories. To recall, language biography interviews as a method originated in research on second-language acquisition; I thus orient to the work of scholars such as Kramsch (2009) and Busch (2006, 2012), who emphasize the importance of biographic accounts for mediating between macro- and micro-level analyses of language. As such, language biographies can open up space for acknowledging the uniqueness and situatedness of people's individual experiences without losing sight of how these individual experiences are connected to broader social discourses. While language biography interviews in this traditional sense are typically used in language learning contexts, I here used the idea of a language biography more loosely to elicit 'folk-linguistic' (cf. Preston, 1996) accounts on professional language use.

I conducted the language biography interviews with the same 21 UX writers I had previously interviewed for the profession mapping (see Chapters 2 and 3). In fact, I often conducted informational interview and language biography interview in the same sitting, moving from questions about UX writing as a whole to more specific questions pertaining to people's personal use of and views on language in their work. As a reminder, my interviewees all worked at medium to large, international software design companies. Most of them were based in the United States, with more than two-thirds of them based specifically in California. The interviews were conducted via video call; with the exception of two cases, I recorded both audio and video. At its core, the language biography part of the interview was organized around questions about how UX writers think about and use language in their own work, with a specific focus on their personal, affective, and/or aesthetic perspectives on language as part of their professional practice. This included, for instance, questions such as: *How were you drawn to this type of work? How would you describe the work that you do? What role do you see language playing in your work? How important is creativity to your work?* The questions thus addressed various aspects of interviewees' professional experiences while centring interviewees' reflexive framing of language in relation to their work. In the current chapter, I focus on one specific part of these language biography interviews, namely creativity. Of course, given the nature of the interviews, people's accounts of creativity in their work often overlapped with their answers to other questions; for instance, my interviewees frequently also touched on what, in their opinion, would count as 'good' UX writing when they talked about creativity.

In terms of processing the interview data, I started with loosely transcribing all interviews. Using the software *Otter.ai*, I first created an automated transcript of each interview; in a second step, I then listened to each interview and manually corrected the transcripts. It is at this point that I also started identifying initial themes in my data; immediately after correcting a transcript, I read through it to annotate participant's answers with short, summarizing statements as a form of 'open' (cf. Glaser, 2016) or descriptive coding. This meant, for instance, annotating instances where interviewees mentioned creativity with a descriptive statement such as 'says that their work is not creative' or even just with a statement such as 'talks about creativity in their work'.

I transcribed my interviews on a rolling basis, which means that as soon as possible after I had conducted an interview, I would transcribe it. Given that I conducted these interviews over a time span of several months, this meant that my initial annotations, too, were sometimes weeks apart. For my analysis, I therefore started with re-reading each transcript and extending my annotations; for this, I imported the transcripts into the qualitative coding software MAXQDA. Using MAXQDA, I then extracted all sections I had annotated as dealing with creativity for a closer, interpretive analysis, whereby I first compared and grouped my annotations into thematic clusters. Through this, I identified three main themes or discourses about creativity: artistic creativity, creativity as problem-solving, and disavowals of creativity, which I here use to offer an in-depth, qualitative, and interpretative analysis of creativity in UX writing.

Alongside the accounts of creativity in the language biography interviews, I also draw on a handful of second-hand, public interviews with UX writers in my analysis. These public interviews were part of the data I had collected during my mapping of the profession (see Chapter 2). Evidently, this data is different from the language biography interviews, as these are instances of UX writers discussing creativity of their own accord, unlike the language biography interviews, where I explicitly asked my interviewees to reflect on the matter. This public discourse is thus complementary to the language biography interviews; I use this data to triangulate but also deepen some of my observations about the way creativity is used as a rhetoric in UX writing.

## 4.3 Analysis: UX Writing and Elite (Non-)Creativity

In my analysis, I discuss three different discourses UX writers use to make sense of creativity in their work: (1) artistic creativity; (2) creativity as problem-solving; and (3) disavowals of creativity. In what follows, I first discuss each of these discourses, before turning to a more in-depth discussion of the third. That (some) UX writers claim that their work is *not* creative may be somewhat surprising, given the positive connotation creativity has both in general and specifically in business and workplace contexts (cf. Wilf, 2019). These moments of disavowal are thus particularly interesting. Indeed, I propose that it is in these moments that the question of what it means to claim creativity as a professional quality or skill comes to the forefront, and it is here that one sees most clearly the power and politics of creativity in language work. I start, however, by discussing the ways in which my interviewees made creativity relevant more generally, for when asked directly, many of them claimed that they were, in fact, creative.

### 4.3.1 Artistic Creativity

In the language biography interviews, I asked my interviewees in a relatively straightforward manner how (and also if) they saw creativity as relevant to their work. In this vein, the focus of my analysis is not on how UX writers' actual, day-to-day work might be considered creative but rather on how they themselves think about and reflect on creativity – on what 'being creative' means to them. Some of my interviewees responded to this prompt by comparing their own work to what scholars have called big-C or historical creativity (e.g. Boden, 2004): the work of artists or poets and the idea of creativity as something new, unique, and aesthetic. As a way of illustrating this, I offer Extract 4.1, where Tina discusses how she uses creativity in her own work. (Tina is once again a pseudonym, as are all the other names I use for my interviewees in this chapter.) At the time of the interview, Tina had been working as a UX writer for more than ten years and was currently team lead at a large, international software company in the United States. What I am particularly interested in here is what creativity means to her and in what ways she sees this as important for her work as a UX writer.

**Extract 4.1: Language biography interview (Tina)**

1. When I was leading the team, it was always vital that we weren't seen as just
2. copying [competitor]. . . . It was always important that we bring a fresh language
3. to it and crea–, you know, and sort of as much as we could to expre–, to express
4. sort of a creative fun side, you know. And that's really where we put a lot of focus
5. on, is sort of fun, engaging, human language.

In this instance, Tina equates creativity with 'express[ing] . . . a creative fun side' (lines 3–4) and with the use of 'fun, engaging, human language' (line 5). In this vein, Tina seems to allude to wordplay, puns, and other poetic forms of language. The three adjectives Tina uses – 'fun, engaging, human' (line 5) – bring to mind the approach to creativity that Carter and his colleagues take, where creativity is understood as a form of language play and said to establish or sustain interpersonal relationships (e.g. Carter & McCarthy, 2004; Atkins & Carter, 2011; see also Norrick, 2016). At the same time, the idea of novelty is underscored here: Tina's main goal is to make sure that her team is '[not] seen as just copying [their competitor]' (lines 1–2); this, she suggests, can be done by using 'a fresh language' (line 2). For Tina, creativity is thus linked to uniqueness and to language that stands out vis-à-vis other kinds of language use.

Overall, Tina's view of creativity is also reminiscent of romantic notions of creativity, where creativity is located both in the new and in the spectacular (Wilf, 2014), which of course stands in contrast to the 'democratized' notion of creativity that sociolinguists such as Carter propose (e.g. Carter,

2004; Maybin & Swann, 2007). Nonetheless, with regards to language, the kind of creativity Tina describes here might best be understood as 'linguistic creativity' – evidently, her focus lies on the creative text or product rather than the creative process. Creativity, here, is language that draws attention to itself, the kind of writing that emphasizes what Carter (2004) calls the 'literariness' of language.

In terms of its function, Tina suggests that the point of using such creative or literary language is to make sure that her (or her team's) work is seen as more than just a copy of their competitor (lines 1–2). This is not just creativity for its own sake; rather, creativity here becomes a differentiating factor, something that makes the writing unique not just – or not even primarily – in a poetic sense, but above all in terms of its business value. The connection between creativity and business value was also mentioned by other UX writers I interviewed. As a second example of this, I offer Extract 4.2, where David talks about creativity and branding. Like Tina, David also works at a large, international software company in the United States (though not the same as Tina's); he equally has more than a decade of experience working with/on language in the UX industry.

**Extract 4.2: Language biography interview (David)**

1. Sometimes we will find a moment when we're like, "Oh, here's a chance for a little
2. branded moment of delight." And then the, the goal becomes the language again.
3. . . . So there are times when the creativity is about the words, is about like how (do)
4. we make this delightful and pleasant. But I think most of the time, it is about can I,
5. if I push myself, can I find a way to explain this or describe this that will be easier
6. to understand for more people.

In this case, David mentions two kinds of creativity, one that is primarily 'about the words' (line 3) and a second that is concerned with making things 'easier to understand' (lines 5–6). The first kind of creativity – and how David initially conceptualizes the matter – is similar to Tina's account. It's a kind of creativity where 'the goal becomes the language' (lines 2). Like Tina, he connects this kind of creativity to business aspects, characterizing such creative language use as 'branded moment[s] of delight' (line 2). Note that there is an ambiguity in what David says; he does not specify who experiences this creativity as 'delightful and pleasant' (line 4). Presumably, it is the people reading these words (i.e. the users), but there are certainly also corporate actors that benefit. What David is creating is, after all, a '*branded* moment of delight' (line 2, emphasis added). Business, creativity, and affect thus become entangled (cf. Thurlow, 2019) – what is linguistically 'delightful' is here presumably used to advance a corporate agenda by strengthening the company's brand. Certainly, the goal is not just to entertain users with linguistic creativity but to make them associate the company with particular, positive values. Here, then, creativity is again both (word)play and implicated in business practices. In both Tina's and David's case, the spectacular aspect of creativity (Wilf, 2014) in their work therefore seems to be tied to their (and their company's) position in a competitive marketplace, whereby creativity, it seems, can be used to stay ahead of others.

At the same time, David suggests that this kind of playful, linguistic creativity that is enmeshed with business and branding is not the only or even main way his work can be creative. As mentioned, David mobilizes two different kinds of creativity in his response; he in fact suggests that in most cases, creativity in his work is not about wordplay but about making things 'easier to understand for more people' (lines 5–6). In this way, he relativizes his claim to linguistic creativity, while at the same time also distancing himself from the corporate goals that seem to be attached to this kind of creativity. Making something comprehensible for users certainly seems more virtuous than making a brand memorable. What begins to emerge here, then, is how creativity is enmeshed also with particular value judgements (cf. Jones, 2018). I will return to this point later; first, however, I elaborate on the second kind of creativity David brings up, the kind that is about making things 'easier to understand' (lines 5–6), or, as other interviewees put it, the kind of creativity that is about solving a problem.

### *4.3.2 Creativity as Problem-Solving*

While some UX writers I interviewed saw their work as creative in terms of linguistic creativity, overall, this sentiment was relatively uncommon. Much more often, my interviewees mentioned this kind of artistic creativity only to then distance themselves from it, emphasizing instead that their creativity was less about the creative use of words and much more about creative problem-solving. To exemplify this, I offer Extract 4.3, where Paula reflects on whether her work as a UX writer might be considered creative or not. Paula has been working as a UX writer for six years. Her lengthy response (slightly shortened in Extract 4.3) suggests again that what counts as creative is not always easy to determine.

**Extract 4.3: Language biography interview (Paula)**

1. I think a lot of it will sort of come down to how you're defining creativity, right,
2. is it, you know, is it sort of creativity in terms of, like, artistic output. . . . If I think
3. about creative writing, like, is it fiction and poetry and storytelling and that or is
4. it thinking about creativity in, like, you know, sort of, like, in the realm of problem-
5. solving, where you're trying to come up with, like, creative solutions. . . . So, I think
6. if we took the first kind of definition of creativity as sort of this artistic output, I
7. would say that there's not a lot of that kind of creativity involved in UX writing. . . .
8. Creative problem-solving is something that I think is a huge part of the role.

What is particularly interesting in this case is how Paula (like David in Extract 4.2) creates an explicit distinction between two different kinds of creativity. On the one hand, she suggests, there is the kind of creativity one might attribute to forms of 'artistic output' (line 2), which she appears to associate with specific genres that are often considered to belong to the domain of literature (as opposed to, say, business writing). This is the kind of creativity one sees in 'fiction and poetry and storytelling' (line 3). Importantly, though, Paula does not see this as the only kind of creativity. She suggests that there is also a different kind of creativity that is 'in the realm of problem-solving' (lines 4–5), and it is this different kind of 'creative problem-solving' (line 8) she then goes on to claim for herself. With regards to her own work, Paula thus understands creativity not as a feature of a specific text or genre but rather locates it in a particular practice. Evidently, this aligns with Jones's (2010) 'discursive creativity', the idea that creativity is not just about the artful use of language, but that there is also creativity in how language is used to do something and to take (social) action. As a result, the focus lies not on a creative product but on creativity as a process. In this sense, the focus of UX writers' creativity is not the text per se but more so the (inter)actions that occur. Once again, this points towards a more-than-representational understanding of communication (cf. Chapter 3).

Across my interviews, claims to creativity as a problem-solving strategy were quite common. Of the 21 people I interviewed, more than half suggested that their creativity was about solving problems rather than artistic use of language. In fact, many of them saw the distinction between linguistic creativity and discursive creativity (or 'artistic creativity' and 'creative problem solving') as connected to different genres, different writing practices, and, ultimately, different kinds of language work. The association with particular genres is partially evident in Extract 4.3, when Paula invokes 'fiction and poetry and storytelling' (line 3) as examples of 'creative writing' (line 3). As said, though, many of my interviewees also suggested that different kinds of creativity corresponded to different kinds of (language) work. Consider Extract 4.4, where Lisa, a senior UX writer with a decade of experience, compares creativity in her work as a UX writer to how someone working in copywriting might be considered creative.

**Extract 4.4: Language biography interview (Lisa)**

1. I wouldn't say, like, I'm– I see it as, like, ((higher pitch)) "I am a creative."
2. ((back to lower/normal pitch)) Like, I don't call myself a creative ((does air
3. quotes)) as a noun. . . . And I think, like, if I were doing work that felt more on the
4. copywriting side of things or the creative writing side of things I would (be), "Yes,
5. I am– my work is creative." But I think this type of work, to me it feels much more
6. like problem-solving and creative in a, in a less, like, artistic way.

Like Paula (Extract 4.3), Lisa creates a distinction between two kinds of creativity: one can be creative in an 'artistic way' (line 6) or by engaging in 'problem-solving' (line 6). Note the lexical similarity in how Lisa and Paula describe these two different creativities; both of them associate one (explicitly) with art and the other (implicitly) with logical and/or critical thinking. Additionally, Lisa here links these different kinds of creativity to different kinds of language work. Artistic creativity is situated in the realm of copywriting, whereas the more analytical creativity is associated with UX writing. Evidently, both conceptualizations are connected to broader discourses of creativity. 'Artistic' creativity, is, as mentioned before, typically entangled with Romantic notions of creativity as something new, unique, and/or spectacular. 'Creativity as problem-solving', on the other hand, is frequently connected to and prioritized in workplace contexts – as Gormley (2020) suggests, the idea of creativity as a problem-solving endeavour and a non-routine or abstract skill is closely connected to neoliberal processes of work and workplace readiness. Given that this is what is favoured in contemporary workplaces, it might not be too surprising, then, that it is this second kind of creativity UX writers tend to claim for themselves when thinking about creativity in the context of their own work.

What is intriguing about Extract 4.4, however, is that even though Lisa ultimately concludes that her work *is* creative in a 'less . . . artistic way' (line 6), she starts by at least partially disavowing creativity. Note how she begins first by voicing an imagined other that *is* a creative, for this, she raises her pitch (line 1) to indicate that while others might say this, she herself would not. She continues by making this position explicit: 'I don't call myself a creative' (line 2), she says. In both instances, she distances herself from the label 'creative'; her use of air quotes after saying 'a creative' (lines 2–3) might also be understood as another distancing strategy.[1] There is, of course, a difference between calling oneself 'creative' and calling oneself 'a creative'. The first can be understood as a general descriptor of anyone and anything, while the latter is typically understood as a specific professional category (e.g. the 'creative class' that Florida, 2012 speaks of). Nonetheless, by saying that she is not 'a creative', Lisa clearly distances herself from a particular understanding of (professional) creativity. This leads me to the third way in which some of my interviewees talked about creativity, namely by disavowing it altogether.

[1] An alternative reading of Lisa's use of air quotes might be that she is using them to underline that she is talking about the noun 'a creative' rather than the adjective 'creative', since she continues by saying that she means this 'as a noun' (line 3). However, given that she overall distances herself from people calling themselves creatives (i.e. the noun), I propose that the air quotes here work as a distancing strategy.

### 4.3.3 *Disavowals of Creativity*

I should preface this by saying that most of my interviewees did suggest their work was in some way or another creative. Nonetheless, there were a few interviewees who explicitly claimed they were *not* creative in their work. This is the case in Extract 4.5, where Terry reflects on the label 'creative' and whether or how it applies to his own work. At the time of our interview, Terry had been working in the tech industry for slightly more than ten years, seven of them in UX writing. What interests me in Extract 4.5 is Terry's explicit disavowal of creativity; at the same time, he is also distancing himself from copywriting (i.e. marketing writing).

**Extract 4.5: Language biography interview (Terry)**

1. When people think of creativity or being creative, they think of trying to be unique
2. or doing something that is, you know, inspiring in some way, and I'm not a
3. marketing writer. And I think there's this sort of joke I've heard a little bit, that UX
4. writing is, like, the boring form of design. That's fine. ((laughs)) Like, I don't care.
5. I'm not– I'm not a marketing writer.

One thing that is of course interesting here is that, like Lisa in Extract 4.4, Terry creates a distinction between professional writing that is more creative (marketing writing in this case) and professional writing that is less creative (UX writing). Such explicit contrasts between 'more' and 'less' creative language work came up multiple times in my interviews, and it is something I will address in more depth in the last part of my analysis. For now, however, I am interested in how Terry is explicitly disavowing the label creative, claiming – relatively proudly – that he prefers to engage in work others might call 'boring' (line 4). Given the positive connotations that creativity generally has, expressing a preference for 'the boring form of design' (lines 4–5) might seem like a strange thing to be proud of, or, perhaps, like a (not so) humble brag; indeed, Terry's laughter on line 4 could be understood as a mitigation of this implicit brag. In this way, then, Terry's disavowal of creativity is not neutral, rather, he is making a particular value judgement, whereby *not* being creative is valued more than those forms of writing that are creative.

Terry also explicitly distances himself from other forms of language work. He twice states, 'I'm not a marketing writer' (lines 2–3 and 5), making a repeated effort to differentiate his own work from that of a marketing writer, which might be more creative, but apparently less desirable. This differentiation of UX writing from marketing and copywriting is common among some (though not all) UX writers, who claim that while UX writing is typically concerned with helping users, copywriting and marketing are primarily about selling things (cf. Chapter 2). In this regard, UX writing is positioned as morally superior. While it is not entirely clear whether Terry is alluding to this distinction, his disavowal of creativity is clearly connected to an elevation of UX writing over other forms of language work.

As said, other interviewees also explicitly disavowed creativity, though sometimes in slightly different ways. Thus, a second theme in these disavowals of creativity was a disavowal – or at least discouragement – of aesthetics, as in Extract 4.6. Here, Scarlett, an established UX writer with more than a decade of experience, explains that some UX writers, especially more junior UX writers, tend to be too focused on aesthetics.

**Extract 4.6: Language biography interview (Scarlett)**

1. I need to not bring my own aesthetic and be like, "This is going to be a work of
2. art." Because it's not. It needs to be designed instead of artistic. And that difference
3. is something that I see– especially people who are brand new to UX writing . . .
4. they come in and they're like, "I'm gonna make it beautiful!" and I'm like, "I don't
5. care about beauty, we need it to work first."

Scarlett does not so much talk about UX writing as something that may be 'boring' (as Terry does in Extract 4.5), but she emphasizes that creativity in the sense of beauty or aesthetics has no place in UX writing. UX writing, she suggests, is not supposed to be 'beautiful' (line 4), rather, it 'need[s] . . . to work first' (lines 5). What exactly 'working' might mean is left implicit here, though the general orientation seems to be towards functionality. 'Form follows function' is a well-established credo in (UX) design. Along those lines, Scarlett indeed establishes a distinction between art and design (line 2), with UX writing, so she suggests, belonging to the latter.

The differentiation between art and design – and the implication that design is *not* concerned with aesthetics – might seem odd at first. The relationship between art and design is in fact complicated, and they are often seen as connected, both historically and in contemporary practice (cf. Walker, 2009). It is important to bear in mind, though, that for UX writers, design is closely associated with *UX design* (rather than, say, fashion design), which, as discussed in Chapter 2, is heavily influenced by computer science and engineering. Design, in this sense, may become shorthand for implicitly – and perhaps not even consciously – affiliating oneself with this lineage (cf. Chapter 3). Design (unlike art) can therefore be associated with functionality and utilitarian value, or, as Scarlett puts it, with making things that 'work first' (line 5). Here, then, Scarlett's disavowal of creativity is also enmeshed with orienting to particular institutional hierarchies: in software design companies, engineers tend to be paid and respected more than designers, who in turn tend to have more pay and more say than writers.

To be clear, I am not suggesting that all or even most UX writers claim that their work is not creative. Indeed, 'creative' is such a positively connoted concept that it is not surprising that most of my interviewees didn't outright disavow creativity but rather suggested that their work was not creative in an artistic sense but might be understood as creative in another sense – recall how Lisa (Extract 4.4) first said that she wouldn't call herself 'a creative' only to then reason that her work does indeed involve creative problem-solving. But it is precisely because creativity is typically seen as something people would rather want to claim than not that I see these moments of disavowing creativity as particularly interesting and intriguing. These moments are worth examining more closely exactly because they are unexpected and might thus reveal something new about the symbolic economy of creativity and/in language work. Concretely, I suggest that these instances of UX writers claiming that their work is *not* creative perfectly illustrate that creativity is indeed not just a linguistic and discursive accomplishment but also a rhetoric. Ultimately, it is in this way that people's claims to creativity (or, in this case, non-creativity) also function as a status claim – which is what I turn to discussing next.

### *4.3.4 (Non-)Creativity as a Status Claim*

So far, I have discussed the metalinguistic claims to creativity that surfaced in the language biography interviews I conducted with UX writers. To complement these explicit metalinguistic reflections, I now turn to my secondary data, that is, data from publicly available interviews with UX writers, to discuss how the notion of (non-)creativity surfaces in such public discourse. Unlike the elicited discourse of the language biography interviews, these are instances where UX writers simply *happen* to discuss creativity in relation to their work. That such public discourse on creativity in UX writing exists is not surprising, given that creativity is a relatively omnipresent paradigm nowadays (Kalin, 2016). What is interesting, though, is that these public interviews again also feature several instances of UX writers suggesting that their own work is not creative, and it is these claims to non-creativity that interest me here.

As a first instance of public discussions of non-creativity, I offer Extract 4.7 from an interview with Roxanne Pinto, who, at the time of the interview, worked as a UX writer for Google. The interview itself was conducted for and published on a blog called *after words*, which outlines career paths for English graduates. In this instance, Pinto explains how language professionals at Google might engage in different kinds of language work and different kinds of writing.

**Extract 4.7: Public interview (Roxanne Pinto)[2]**

1. If you want to do creative writing, write things that aren't as functional in nature,
2. like ads or website copy. That's more about grabbing attention and being

[2] Source: *Roxanne Pinto: Making tech sound more human.* Retrieved from https://afterwordsproject.com/2020/04/28/roxanne-pinto/ (accessed 4 May 2022).

3. entertaining. Or you could develop technical content, which is writing about how
4. things work for the people building it . . . . If you want to do short-form writing,
5. that's microcopy—also known as UX writing. That copy is super short and super
6. functional.

What is striking here is the way that Pinto distinguishes between different kinds of language work and that she explains these differences in terms of how creative the writing is. Like in my interviews, then, the rhetoric of creativity is here mobilized as a means of producing boundaries. Equally familiar is how Pinto notes that among these different kinds of language work, UX writing is not creative but instrumental; it is the copy that 'is super short and super functional' (lines 5–6). In contrast to this, Pinto conceptualizes creative writing as writing that is 'entertaining' (line 3) and 'grab[s] attention' (line 2), presumably by showing features of literariness (cf. Carter, 2004). Through this, Pinto seems to orient to linguistic creativity or to what some of my interviewees (Extracts 4.3 and 4.4) dubbed 'artistic creativity' and not the kind of creative problem-solving or discursive creativity most of my interviewees claimed for themselves.

At the same time, Pinto constructs creative and functional writing as something that is mutually exclusive: one can either write things that are creative but not functional, or functional but not creative; with UX writing, it seems, as the ultimate form of functional writing. There are certainly also judgements at work here about what kinds of functions language is supposed to fulfil in these different cases. Note however that Pinto seems to make no explicit value judgement about whether 'creative' or 'functional' writing is better. Rather, both are constructed as having value, and whether one ends up writing creatively or not is said to be a matter of personal choice, of what 'you want to do' (line 1). Still, there is something intriguing about Pinto negating the label 'creative' for describing her *own* work. After all, people supposedly 'prefer to be creative rather than not' (Toolan, 2012, p. 19). The question is thus what Pinto might gain by claiming to *not* be creative. I propose that in doing so, Pinto can use creativity to set up the value of functionality. In this way, the value of UX writing can be constructed against the foil of creativity and gains its value through its contrast with other kinds of (good) writing.

Of course, not all discussions of non-creativity in UX writing paint creative forms of writing in a positive light. As is evident in some of the interview extracts I discussed above, sometimes, being creative – that is, being 'artistic' – is framed as less desirable and perhaps less prestigious than abstaining from such artistic creativity. This is also the case in Extract 4.8, which shows a short discussion between two UX writers, Ryan Farrell and Patrick Stafford, on the podcast *Writers of Silicon Valley*. This podcast is aimed primarily at other UX writers and consists of episodes where one UX writer (Stafford) interviews other UX writers about their work and their professional trajectories. Again, the topic of the discussion is not creativity per se, but it is something that comes up as these two UX writers talk about what constitutes good UX writing and whether there is such a thing as UX writing best practices.

**Extract 4.8: Public interview (Ryan Farrell and Patrick Stafford)**[3]

**Farrell** I mean, even if you're not necessarily a good writer, you can still be a UX writer that is a very (.) functional designer of words, as long as you understand that it's the user first and, and it's the, it's– I mean, it's the word second ((laughs)) [(to) be perfectly honest that=]
**Stafford** [exactly (.) yeah ]
**Farrell** = that's exactly what it is
**Stafford** creativity is overrated [uhm]
**Farrell** [– it ] really is and I don't think (.) even you have to be that creative to be a UX writer. I don't. I really – I honestly don't think that you do=what I think you just have to have (.) the capacity to put your own ego aside and just y– make it all about em– empathy and empathizing for what t– the product needs to do as you're, as you're producing it (.) and making it come to life quote unquote ((laughs))
**Stafford** absolutely (.) I totally agree

Here, two UX writers emphatically agree that 'creativity is overrated' (line 7). Instead, what is considered valuable is a UX writer's capacity to understand the context of the software and what it is supposed to do. As with Pinto's statement in the previous extract, functionality is once again framed as a positive opposite to more creative forms of language use. In fact, in this case, functionality is conceptualized as better or more desirable than creativity. A UX writer, as a 'a very functional designer of words' (lines 2), puts their own ego aside (line 11), does not indulge in creativity, and knows that 'it's the user first and . . . the word second' (lines 3–4). Note also the orientation to design again in Farrell's description of a UX writer as a 'designer of words' (line 2). Creativity is here conceived of as something that is connected to a person's ego: creativity not only draws attention to the words themselves but also to the person who wrote them, which, Stafford and Farrell say, is not the point of UX writing. In fact, as I will discuss in Chapter 5, UX writers typically pride themselves on making the words they write as unobtrusive and 'invisible' as possible.

[3] Source: *Episode #6 - Good UX writing is boring, with Ryan Farrell* (20 July 2019). Audio and full transcript by Writers of Silicon Valley available here: www.writersofsiliconvalley.com/episodes/2019/7/20/episode-6-good-ux-writing-is-boring-with-ryan-farrell (accessed 4 May 2022).

Like Pinto, then, these two UX writers suggest that UX writing should be about function rather than form; there is, therefore, no room for wordplay and similar kinds of linguistic creativity. Arguably, this distinction between different kinds of writing is connected to a separation of form and function: creative writing is (supposedly) all about linguistic form, whereas UX writing is supposed to be all about the function of these words (cf. Chapter 3). From a sociolinguistic standpoint, such a distinction between form and function is somewhat arbitrary and perhaps even suspicious. While sociocultural linguists might also sometimes separate the two for analytical purposes, they tend to agree that language always entails form and function (cf. Jakobson, 1960) and that communicative practices can ultimately only fully be explained by considering both (see, e.g., Lillis, 2013, ch. 3). As such, the assessments of these UX writers about the functionality of their own writing are based on a misrecognition of how language works.

The discussion the two UX writers in Extract 4.8 have on functionality versus creativity also highlights a second point, namely, that what kind of writing is considered – and supposed to be – creative appears to be very much connected to what that writing is meant to do. Like Scarlett in Extract 4.6, Farrell and Stafford also imply that UX writing should be functional rather than simply aesthetic or expressive. This way, the discussion in Extract 4.8 is also a discussion about semiotic ideologies (Keane, 2018), that is, about the meaning and function of (creative) signs.[4] Because presumably, Farrell and Stafford do not mean to suggest that *all* creativity is overrated, but rather that it has no place in the professional practice of UX writing. In an interesting twist, then, the emphasis on form and symbolic value one often sees with regards to language work in late capitalism (cf. Duchêne & Heller, 2012; see also Kelly-Holmes, 2020) is reversed: here, form is precisely not the point. Importantly, though, this reversal comes from a deliberate and reflexive position; to be creative or not in this case constitutes a choice.

As a matter of choice, disavowing creativity is neither a passive act nor a neutral one: non-creativity is evidently connected to privilege and status. As a way of discussing this, I return to my own interviews with UX writers and offer Extract 4.9, where Melanie reflects on when one should and should not be creative as a UX writer. This short statement nicely encapsulates the point – and the problematics – of claiming non-creativity. At the time of our interview, Melanie had been working as a UX writer for two years. While she was less experienced than some of the other UX writers I interviewed, her perspective

[4] I deliberately use the concept of semiotic ideology here rather than linguistic ideology in order to acknowledge that language work – especially that of UX writers – often also involves modes other than language. I offer a more detailed discussion of both concepts in Chapter 5.

on creativity was consistent with the views of most of my interviewees – that UX writing wasn't necessarily creative in an artistic sense, but that it could be considered creative in different ways.

**Extract 4.9: Language biography interview (Melanie)**

1. I think th– you can be very creative as a UX writer, but you also have to be very
2. pragmatic and know when not to be creative or you have to understand the place
3. of creativity in UX writing.

Likely owing to the fact that she is explicitly asked to think about creativity and her work, Melanie is somewhat less extreme in her disavowal of creativity than Farrell and Stafford (Extract 4.8). She doesn't call it overrated; indeed, she even says that one 'can be very creative as a UX writer' (line 1). Importantly, though, and this is where I suggest one sees the same discourse as in the two public interviews discussed before (Extracts 4.7 and 4.8), Melanie suggests that the important thing in UX writing is 'know[ing] when *not* to be creative' (line 2, emphasis added). What distinguishes a good UX writer, then, is their ability to reflect on when to be creative and when not to be. Again, creativity becomes a matter of choice – and this, I suggest, is the point of these claims to non-creativity, or this is how they turn into a status claim: framed as a choice, the claim to non-creativity becomes an expression of mastery of both creative *and* non-creative writing. The notion of creativity and creative writing then becomes a foil for UX writers in two different but connected ways: first, as something against which the virtues of 'functionality' can be constructed, and second, as something that a (good) UX writer can consciously choose to do or not do.

Evidently, such claims to non-creativity are not neutral. As Jones (2018) reminds us, saying that something is creative means making a value judgement. Likewise, saying that something is *not* creative is equally enmeshed with judgements about the symbolic – and potentially material – worth of something. In the case of UX writing, these claims to non-creativity can be understood as a form of symbolic negation (Bourdieu, 1991), whereby relatively privileged language workers characterize their own writing as non-creative under the premise that this is not for lack of skill but owing to a deliberate choice. Presumably, other kinds of language workers, specifically less elite language workers, could not make the same status claim when describing *their* work as not creative. To put it more bluntly: claiming that one does not care about creativity and would rather engage in boring work is indeed a humble brag, because such symbolic negation or condescension (Bourdieu, 1991) is, of course, only possible from the position of the (more) privileged (see also Mapes, 2021). It is in this way that claiming to be *not* creative is based on a fundamental privilege of disavowal.

## 4.4 Conclusion: The Symbolic Economy of (Non-)Creativity

> All particular linguistic transactions depend on the structure of the linguistic field, which is itself a particular expression of the structure of the power relations between the groups possessing the corresponding competences. (Bourdieu, 1977, p. 646)

Nearly five decades ago, Bourdieu (1977), pointed to the many ways in which linguistic exchanges operate under the logics of a marketplace. Ways of speaking, he argued, gain their value not based on intrinsic properties but based on their relation to other ways of speaking and on the power relations that structure and govern what he calls the linguistic marketplace. In other words, the symbolic and economic value of language is determined by its structured and ideological basis. As I have argued in this chapter, creativity, too, is determined by semiotic ideologies (cf. Keane, 2003, 2018), and therefore depends on people's assumptions about what forms of speaking/writing can and should be creative – and on their understanding of which people can and should be creative.

In this chapter, I have also shown that there is no singular way in which UX writers understand creativity in their work. Some see their work as linguistically creative (cf. Carter, 2004), although most emphasize that their creativity lies rather in 'creative problem-solving' – that is, discursive creativity (cf. Jones, 2010) – and in finding solutions to particular (business) questions. Additionally, I've suggested that such a discursive approach to creativity is connected to UX writers' more-than-representational – or posthumanist – understanding of language. (I will return to this in Chapter 6, where I address the consequences of this perspective on creativity and writing through a discussion of particular textual examples of UX writing.)

Beyond the (re-)framing of creativity as problem-solving, some UX writers also claim that their work is not creative. One thing that becomes apparent through these different opinions is that 'creativity' itself is not a fixed term, but rather something that people repeatedly and discursively construct. In this sense, creativity is not simply or straightforwardly a matter of the linguistic and discursive forms that it can take, but also a matter of what people come to claim creative writing or speaking to be in the first place. In other words, creativity is also a discursive strategy or rhetoric people can draw on in their constructions of social reality (cf. Thurlow, 2019). This shifts the question from 'What is creativity?' and 'Why is *this* creative?' to 'Why does it matter that someone calls this creative?' Understood thus, creativity is not something that occurs *in* discourse, but something that is a discursive phenomenon in and of itself. Discussions over what makes people creative do not become irrelevant, however, their point changes: it is no longer about pinning down creativity, but about understanding the logics of its power and of understanding how creativity is 'located in wider systems of privilege and inequality' (Thurlow, 2019, p. 101). It is in this vein that looking at the rhetorics of creativity that are at work in UX writers' statements can tell researchers something about what *counts* as creative language.

The distinctions UX writers draw between creative and non-creative writing and the claims to (non-)creativity that are enmeshed with this could help sociocultural linguists better understand how creativity as a rhetoric can be mobilized to construct and assert hierarchies of language work. To quote Bourdieu (1977) once again:

> It is they who, in the certainty that they incarnate the linguistic norm, can permit themselves transgressions which are a way of affirming their mastery of the norm and their distance from those who blindly adhere to it. (p. 659)

As I have suggested, claiming non-creativity is one way for UX writers to discursively construct their professional identity in a positive light. Building on the observations of (non-)creativity made in this chapter, I suggest theorizing creativity not as something that can be defined, rather, my point is that the nature of creativity depends on the particular circumstances in which discourses of creativity are deployed. In this way, then, creativity surfaces as something that is emergent and dependent on particular material-discursive enactments. Consequently, the conditions under which it is professionally valued to be seen as creative (or not, as I have argued in this chapter) depend on the particular discourses that structure the professional field in any given moment in time. From the posthumanist perspective that this book is advancing, one might say that (non-)creativity is thus constituted through particular intra-actions that produce something as creative or not. In this way, it is through a complex configuration of agential cuts that UX writers' non-creativity is produced as valuable.

Concretely, I suggest that UX writers' ability to turn non-creativity into something positive is precisely because of their relatively secure and prestigious position in the hierarchy of language workers; this grants them the privilege of disavowing creativity. As people who have mastered the 'norm' of writing creatively but who choose to not do so, they can permit themselves to transgress from the norm while still making a status claim (cf. Bourdieu, quoted above) – something that other, less privileged language workers cannot do. (Non-)creativity is then not only a means to claim status but something that is thoroughly enmeshed with structures of power and inequality. Either way, it appears that what counts as creative language – and what not – is an important concern for UX writers both in delimiting their domain of language work and in negotiating their relative position in the hierarchy of language workers.

Discussions of creativity have in the past often excluded the realms of business and professional practice. Recently, however, there has been increased interest – both academic and popular – in creativity and innovation in business contexts (Wilf, 2019). In contemporary economic development and policy discourse, creativity is often characterized as something that can be mobilized to transform economies and communities and is linked to economic growth (Foord, 2009). A case in point is Florida's (2012) notion of the 'creative class', which he sees as the driving force of the economy. As Gormley (2020) shows, similar discourses of economic growth also permeate educational policies, where creativity is often co-opted into neo-liberal themes and processes of financialization that prioritize workplace readiness and employability over other outcomes. From a workplace or business perspective, then, creativity typically has high cultural and professional value; there is an overall positive valuation of creativity as competitive advantage.

While creativity has gained renewed attention in certain academic discourses, the topic of professional creativity in language work still remains relatively overlooked. Unjustly so, for much of what Florida (2012) characterizes as 'creative work' is inherently tied to linguistic or semiotic work; at the same time, and as I have discussed in the previous two chapters, language is also increasingly seen as a source of competitive advantage in contemporary economies. An exception in this regard is the work of Bhatia (e.g. 2012, 2016), who discusses intertextual and interdiscursive creativity in corporate annual reports, and work by Jones et al. (2012), who examine creative collaboration in the public relations industry. But while Bhatia's and Jones's research indeed examines the creativity of specific language workers (PR writers, in both cases), they both focus primarily on the *shape* that creative practices in language work may take. What remains unaddressed are the views of such professional language workers on what counts as creativity and on why or how creativity might matter for their work. Yet, it is exactly such understandings of how creativity is valued, I suggest, that allow scholars to better understand how discourses of creativity also contribute to structuring the hierarchies and economies of contemporary language work.

Just as the value of creativity usually depends as much on the creator as it does on the creative object, the recognition of language work, too, depends not just on the linguistic work but also on the worker and on where the work is being done. And just as language work is entangled with people's beliefs about language(s), creativity, as I have suggested in this chapter, also depends on people's views of what creative signs are and how such signs are supposed to be used. In other words, creativity, too, depends on semiotic ideologies (Keane 2003, 2018), that is, on the particular meanings and functions people attribute to signs. This entails what forms of communication may count as creative in the first place, but also where and by whom these creative forms of communication may be used – whether, for instance, creativity is deemed acceptable in particular forms of (language) work. In this regard, what emerged in the current chapter is that UX writers typically understand their work as being concerned with the function of words more so than their form. Staying with the topic of semiotic (and media) ideologies, I explore this aspect further in Chapter 5, where I examine how UX writers strive to write interface texts in such a way that they (ideally) become invisible. It is at this point that I turn also to the cultural-political consequences of UX writers' work; the next two analytic chapters thus make up the third and last part of this book, where I am specifically concerned with how UX writers' understanding of language and communication impact their text production in normative ways.

*III*

# Producing Little Texts

*Politics and Power in UX Writing*

# 5 (De)Constructing the Invisible Interface

## Semiotic and Media Ideologies in UX Writers' Work

In Chapter 4, I have considered how UX writers often – and sometimes quite vehemently – emphasize that their language work should not draw attention to itself. This claim to *non*-creativity, I have argued, allows UX writers to negotiate their position in the wider hierarchies of language work. Building on this, the current chapter addresses how such inconspicuousness – or invisibility – is not only a matter of status production but also centrally connected to the production of social and cultural norms. Specifically, I focus on the ideology of the invisible interface and how it both impacts UX writers' status as language workers and has normative consequences in terms of how their work impacts users of digital media.

That invisibility is central to the work of UX writers is not surprising. Indeed, in his definition of wordsmiths, Thurlow (2020c) suggests that invisibility might be a defining feature of most elite language work. The success of most wordsmiths' work, says Thurlow, hinges on its backstage nature: wordsmiths are typically unseen by the wider public; owing to this, their influence is often misrecognized and obscured. As such, the work of wordsmiths is characterized by complex, intersecting hierarchies of invisibility and eliteness, as I seek to illustrate in this chapter. UX writers seem to be a perfect case in point. Their work is invisibilized in such a way that most users of digital media are not aware that this professional role even exists; more than this, UX writers themselves typically emphasize that the interface texts they produce *should* be invisible. Ubiquitous as the work of UX writers may be, it is also (ideally) fleeting, unremarkable, and inconspicuous. Yet, the point I want to make in this chapter is that it is exactly this inconspicuousness that makes UX writers' work influential in potentially normative ways. The invisibility UX writers strive for evidently hinges on privileging a particular kind of seeing, a particular position from which things are rendered visible (or not). Following Thurlow's (2020c) observations, I am thus interested in invisibility both in terms of the status that it affords language workers and the power that can be exerted through it.

My aim in this chapter is two-fold. First, I am interested in how UX writers understand and negotiate the (in)visibility of language in their own work.

Tracing the semiotic ideologies (Keane, 2003, 2018) of these professional language workers, I discuss how UX writers operationalize a discursive ideal of invisible writing in order to establish the value of their linguistic work vis-à-vis their colleagues, who typically privilege other modes of meaning making. Second, I examine how UX writers make sense of the linguistic and cultural-political (or ideological) consequences of this invisibility. In this regard, I suggest that the ideal of the invisible interface is a central media ideology (Gershon, 2010) that not only structures the work of UX writers but ultimately determines how ordinary users can(not) communicate with and through digital media. With these two objectives in mind, I first situate my research vis-à-vis scholarship on (in)visibility in language work and discuss differences between the invisibilization of low- and high-end language work. Next, I introduce the main theoretical concepts framing my case study: semiotic ideology and media ideology. In the main part of this chapter, I then turn to examining the interplay of invisibility, status, and power in UX writing. For my analysis, I draw on 21 language biography interviews with UX writers (see also Chapter 4) and on a small collection of 33 'best practice' blog articles written by UX writers. I first discuss how UX writers mobilize the (in)visibility of their work to claim linguistic expertise, before examining invisibility as a key media ideology through which UX writers construct the interface as a neutral, apolitical carrier of information. I end by linking my case study to broader discussions of invisibility in cultural studies of technology, arguing that communication with and through digital media is shaped not only by users' perspectives but also by the semiotic and media ideologies of its producers.

## 5.1 Thematic Context: (In)Visibility and Language Work

While language has become pervasive in most domains of work, it often remains invisible – even in those professions where language is the main objective and output. This may be because language is taken for granted (for instance, in service work; see, e.g., Duchêne, 2011; Lønsmann & Kraft, 2017); at other times, though, language work can also be purposefully obscured, as is often the case in elite language work (see Thurlow, 2020c). As such, the linguistic labour of both low-status and elite language workers can be invisibilized. What differs, however, is the conditions under which this invisibilization occurs and the consequences that it has.

In the case of wordsmiths, invisibility is often seen as self-chosen and/or a matter of professional skill (and thus prestige; cf. Chapter 3). A typical example of invisible yet prestigious language workers are the speechwriters that Mapes (2025) discusses in her work. These markedly high-end language workers are typically the *authors* but never the *animators* or *principals* of their own work (cf. Goffman, 1981, on production roles). Yet, as Mapes (2025) demonstrates, in the case of speechwriters, the strategic erasure of their involvement ultimately becomes a resource for their claims to professionalism in the larger professional and sociopolitical context of their work. Droz-dit-Busset (forthcoming) likewise shows that copywriters take pride in disappearing behind their writing; consumers may see their words, but never copywriters themselves. Here, then, invisibility is a necessary but at the same time explicitly valued aspect of wordsmiths' work. In contrast to this, the invisibility of low-status language workers is arguably less voluntary. For instance, in his work on the linguistic labour of employees in an airport passenger- and baggage-handling company, Duchêne (2011) shows that employees become visible only if and when the company can make use of their linguistic services and that they are relegated to the background as soon as that usefulness has passed. In other words, low-status (language) workers' invisibility is a matter of their being excluded from the social gaze (cf. Star & Strauss, 1999; Arrivé, 2020). This kind of invisibility is not managed by the language workers themselves; rather, it is dictated by the needs (or desire) of others.

Though this lies outside the scope of the current chapter, I should note that the deliberate construction of invisibility is also a central organizing principle of the labour market of the software and technology industry UX writers are situated in. Work by critical media scholars such as Irani and Silberman (2013), Roberts (2019), and Gray and Suri (2019) powerfully documents how the prestigious work of software engineers is dependent on the hidden manual labour of low-paid workers taking care of content moderation, manual transcription, image checking, and so on. The chosen invisibility of high-status workers is made possible by the enforced invisibility of others. There is a simple point that I wish to illustrate with this short comparison of invisibility in low-status and elite (language) work: no work is inherently visible or invisible; rather, (in)visibility is always relational, whereby some things are foregrounded while others are not (cf. Star & Strauss, 1999). What follows from this is that the value of (in)visibility is not something that can be determined a priori; rather, it is something that is established through discourse and practice. This may seem like a banal observation, but it is central to my argument: that invisibility can be a resource for establishing status and prestige.

As a resource for status production, invisibility in language work is inevitably connected to questions of hierarchy and power. More than this, it is also a matter of normativity. In this regard, scholarship in translation studies proves to be particularly insightful. Here, invisibility and its sociocultural consequences have been heavily discussed and theorized. (For an overview, see Koskinen, 2000.) Views on invisibility in translation studies are influenced most prominently by the work of Venuti (1995), who, writing in 1995, offered a powerful critique of the 'illusion of transparency' that is created through translation (pp. 1–42). Pointing to the ethnocentrism inherent in a so-called transparent translation, Venuti argues that invisibility in translation is never neutral but rather a highly normalizing practice, whereby texts are assimilated into a target culture. The field of translation studies has since engaged with (in)visibility as a moral (or ethical) matter (Koskinen, 2000). For my purposes, however, the case of translation – as another form of elite language work (see Jaffe, 2020) – shows one thing clearly: that (in)visibility is also tied to people's understanding of how a text *should* work. This in turn is central to understanding invisibility in the context of UX writing, for the 'illusion of transparency' (Venuti, 1995) that has shaped (and, arguably, still shapes) folk understandings of what constitutes a 'good' translation is strikingly similar to the way UX writers talk about invisibility, all of which is connected to their understanding of how language in digital media interfaces as well as digital media interfaces in general are supposed to work. This brings me to a short introduction of the theoretical concepts I use in my analysis in this chapter: semiotic and media ideologies.

## 5.2 Theoretical Framing: Semiotic and Media Ideologies in UX Writing

One of the most pervasive notions about what constitutes good UX writing is that interface texts are supposed to be unnoticeable for users of digital media. As I will show later, this ideology has its roots in UX design and human–computer interaction (HCI), where the invisible interface is a longstanding ideal that is still perpetuated today (Pucheu, 2016; Petrick, 2020). For now, however, I want to outline how this is centrally a matter of semiotic and media ideologies, both of which are essentially concerned with people's beliefs about how communication works.

Semiotic and media ideologies both draw on earlier work carried out in linguistic anthropology on language ideologies (e.g. Silverstein, 1979; Irvine, 1989; Woolard & Schieffelin, 1994). Generally, scholarship on language ideologies concerns itself with how language is connected to cultural and social values and how these connections are articulated through language. Language ideologies are people's rationalizations or justifications of (perceived) language use (Silverstein, 1979); this includes beliefs about who should speak, how one should speak, or, even simpler, what language one should speak (see, e.g., Schieffelin et al., 1998). Crucially, then, these beliefs are never only about language or communication, they always also entail judgements about how specific forms of communication are connected (or seen as connected) to broader social discourses. In other words, beliefs about language always also entail normative beliefs about the social and moral value of particular users of language. In this way, the concept of language ideology entails a particular emphasis on reflexivity as well as power.

Language ideologies are typically seen as pertaining to people's normative understandings of *language*. Communication, however, is always multimodal, and UX writers' views on language are, arguably, never just about language; they always also entail judgments about other modes and about media more generally. Given the highly multimodal and necessarily digitally mediated nature of UX writing, it is thus useful to also draw on the related concepts of semiotic ideology (Keane, 2003, 2018) and media ideology (Gershon, 2010). To this end, and following Thurlow (2017), I understand language, media, and semiotic ideologies as closely interrelated. As Thurlow points out, separating them is often done more for analytical convenience rather than to do their on-the-ground effects justice (see also Gershon, 2010). It is for the sake of analytic precision that I discuss semiotic and media ideologies separately in this chapter; I return to their inter-connectedness in my discussion and conclusion.

UX writers' work is always multimodal (cf. Chapter 2); to address how this impacts their understanding of communication, I draw on Keane's (2003, 2018) concept of semiotic ideology. While semiotic ideology is sometimes understood as an umbrella concept encompassing any ideologies about communication, I here understand the term more narrowly as pertaining to people's beliefs about specific sign systems and the differences between them. Understood thus, semiotic ideologies enable scholars to focus on the meanings and functions people attribute to various modes. This entails, for instance, beliefs about the epistemological status of different modes – whether, for instance, an image is more 'truthful' than spoken or written language (cf. Ehrlich, 2019; see also Thurlow, 2017). As such, semiotic ideologies are about people's assumptions about what different forms of communication can and should do. Like language ideologies, these beliefs about different forms of communication – or different kinds of signs, as Keane (2018) puts it – are entangled with broader social, cultural, and political processes. This, I propose, is useful for understanding how UX writers value the role of their linguistic work vis-à-vis other semiotic work that is done in software design (most notably, of course, visual design), all of which is connected to the sometimes voluntary and sometimes involuntary invisibility of UX writing.

At the same time, UX writers' work is centrally shaped by its digitally mediated nature; therefore, I turn to Gershon's (2010) concept of media ideology. As the name suggests, the concept of media ideology draws scholars' attention to *mediality* as a central meaning-making resource. Media ideologies thus emphasize 'how people understand both the communicative possibilities and the material limitations of a specific channel, and how they conceive of channels in general' (Gershon, 2010, p. 283). This entails judgements about the social functions, meanings, and values people ascribe to media choices, such as whether writing an email or making a phone call is deemed more appropriate for a particular task. More implicitly, media ideologies are also at work in people's understandings of how communication is *mediated.* This includes, for instance, beliefs about whether face-to-face interaction is more 'immediate' than interaction via video call; this is what Bolter and Grusin (1999) call the 'logic of transparent immediacy', which structures much contemporary conceptualizations of media and mediality (pp. 20–31). It is this second aspect that is particularly helpful for understanding media ideologies in UX writing and for examining how UX writers conceive of software interface texts as transparent (in the sense of Bolter and Grusin), thereby erasing the mediated nature of their own work. This, in turn, has consequences not just for UX writers but also for users of digital media, as I discuss in the case study that follows.

## 5.3 The Data

In order to examine the semiotic and media ideologies that underpin how UX writers negotiate invisibility in their work, I draw on two complementary datasets I collected during my research:

- 21 language biography interviews, which constitute a reflexive and folk-linguistic account of UX writers' language expertise
- 33 'best practice' blog articles on UX writing; this data gives insight into ideologies that are circulated more widely within the professional field of UX writing

I offered a detailed overview of my interview procedure in Chapter 4; but as a reminder, these were semi-structured interviews I conducted via video call with 21 UX writers from large, international companies, most of whom were based in the United States. The interview questions aimed to elicit 'folk-linguistic' (cf. Preston, 1996) accounts of UX writers' personal and affective perspectives on language in their work. After loosely transcribing the interview data, I conducted a first round of descriptive coding during which I annotated participants' answers with short, summarizing statements in order to identify recurrent themes in their answers. While invisibility was not something I actively inquired about in these language biography interviews, my initial thematic coding of the interviews revealed that the idea of 'invisible' writing was something that slightly more than half of my interviewees (12, to be precise) mentioned when talking about what they considered good UX writing. This, I felt, made it a recurrent enough topic to warrant special attention. Thus, unlike UX writers' views on creativity, which I discussed in Chapter 4 as a form of explicitly elicited metadiscourse, my concern here lies with more general and at times more implicit ideologies structuring UX writing – the kinds of issues that UX writers themselves make relevant when describing their work.

As with Chapter 4, I supplemented the language biography interviews with data I collected during the profession mapping phase of my research. In this instance, I turned to 33 'best practice' blog articles on UX writing. These are texts of medium length (ca. 2000–4000 words) in which UX writers describe the nature and goals of their work, typically by way of explaining what UX writing is and by giving examples of what counts as good or successful UX writing. I collected these blog articles by archiving articles that had been featured in two key industry newsletters about UX writing, *The Dash* (published by the *UX Content Collective*) and *UX Writing Weekly* (published by the *UX Writing Hub*), as well as articles that had been posted in LinkedIn groups about UX writing that I had joined. As such, these were all texts that had been given at least some prominence within the professional community of UX writers. By combining these two datasets, I seek to illustrate how invisibility structures the work of UX writers across various practices, ranging from implicit metalinguistic reflection and public professionalization tactics (cf. Chapter 3) in the blog articles to more explicit metalinguistic reflection and accounts of their day-to-day work in the language biography interviews. In this sense, my insights in this chapter are based on a close reading of two types of interconnected data, across which I identify invisibility as a prominent theme that is central to UX writers' (elite) professional status as well as their larger cultural-political impact.

For the analysis that I discuss here, I started by importing both datasets (i.e. interview transcripts and archived rich-text versions of the blog articles) together into the qualitative coding software MAXQDA. Having worked with the language biography interviews and the 'best practice' blog articles at various earlier stages in my research, I already had a good overview of both datasets. Again, I had specifically identified – and partially coded – invisibility as a topic of interest in my interview data, though only with regards to interviewees' answers to questions about good UX writing. For the current analysis, I re-read each transcript to identify other instances where invisibility was invoked. The blog articles had not been coded before; here too, I started by identifying any instances that discussed invisibility. For both datasets, this meant that I annotated those cases where UX writers mentioned the words 'invisible' or 'invisibility', but also those cases where they indexed invisibility by, for instance, talking about 'hiding', '(not) noticing', 'obscuring', and so on. Additionally, I also annotated instances where UX writers mentioned or indexed (heightened) visibility. In my close reading of the extracts, I identified two main themes: (a) invisibility as a resource for value-creation and (b) invisibility as implicated in power and normativity. My analysis in this chapter is organized around an interpretive and critical discussion of these two themes, which I examine, as said, through the lens of semiotic ideology and media ideology. While I keep semiotic and media ideologies relatively separate in my analysis, they are, of course, connected; I return to this point in my discussion and conclusion.

## 5.4 Analysis: Negotiating (In)Visibility in UX Writing

### *5.4.1 Constructing Value: Making Words (In)Visible*

When it comes to status production in UX writing, visibility and invisibility are closely intertwined. While UX writers may strive to produce writing that goes unnoticed by users, at the same time, they have to ensure that they do not erase their own visibility as (language) workers – at least not within their professional field. Indeed, such professional visibility is a central contention for most UX writers. As an entry point to these matters of status and visibility, I offer Extract 5.1, taken from a blog article titled *Getting a seat at the table as a UX writer.*[1] Here, a UX writer from Dropbox lists ten tips for establishing the importance of UX writing within a company.

**Extract 5.1: Visibility in 'best practice' article**

1. Tip 6: Show the value of UX writing . . . By explaining your decisions objectively
2. and on a regular basis, you'll elevate your visibility as the go-to words person.

As part of these tips, readers are emphatically told that to gain influence and authority, UX writers need to 'explain[] [their] decisions objectively' (line 1) so as to 'elevate [their] visibility as the go-to words person' (line 2). In other words, by making overt their metalinguistic knowledge, UX writers can draw attention to their linguistic expertise and establish the (economic) 'value' (line 1) of their work. Visibilizing words – or, rather, the production of words – thus seems to be part and parcel of being a UX writer. It is also, I suggest, a matter of de-naturalizing the linguistic labour of UX writers, of framing their writing as a skill they have acquired through training and education rather than as an innate quality that all people possess (cf. Duchêne, 2020). This overt display of metalinguistic knowledge constitutes an 'appeal to professionalism' (Fournier, 1999) in order to claim a particular position of status and expertise. I will return to this point later; first, however, I want to unpack how UX writers engage in such visibility work and how this relates to issues of semiotic ideology. For this, I turn to my interview data.

With regards to the potential invisibilization of their work, many of my interviewees suggested that while most people working in the web and software design industry would generally accept (and expect) that the visual

[1] Source: *Getting a seat at the table as a UX writer*, Retrieved from https://medium.com/dropbox-design/getting-a-seat-at-the-table-as-a-ux-writer-da63303d5b1d (accessed 10 October 2022).

design of software had to be created by an expert, the verbal or linguistic aspects of the interface would still be treated as an afterthought, as something that could potentially be written by anyone. Clearly, this is a matter of semiotic ideology: language is considered not only less important than the visual mode but also seen as somehow less complex, as something that anyone – or at least anyone with the credentials to work at a software company – can produce. Therefore, the status of UX writers' work vis-à-vis that of their colleagues seems to hinge on UX writers being able to successfully push back against an (over-)privileging of the visual mode in software design. This was particularly apparent in my interview with Lisa,[2] who had been working as a UX writer for slightly more than a decade when I interviewed her. In Extract 5.2, she talks about her struggle in asserting the importance of her work, suggesting that her colleagues – especially those working on the visual design of interfaces – typically do not understand the importance of words in user interfaces.

**Extract 5.2: (In)visibility in language biography interview (Lisa)**

Communication can happen in a bunch of different ways. Visual communication
is one way, like, the sequence in which you present information is important. But
then the actual words you use are super important. And I think somewhere along
the way, the design field lost sight of that. . . . It's like they completely forgot that
words are part of the picture. ((laughs))

Evidently, Lisa feels that her skills are invisibilized: her colleagues have all but 'forgot[ten]' (line 4) that the verbal design of software interfaces matters. In contrast to this, she positions herself as someone who understands that communication is multimodal, or as Lisa puts it, that 'communication can happen in a bunch of different ways' (line 1), with both 'visual communication' (line 1) and 'the actual words' (line 3) playing an important role. Interestingly, Lisa seems to understand visual communication less in terms of *representation* and more so in terms of the spatial or even temporal *composition* of a text. It is 'the sequence in which you present information' (line 2) that matters, which suggests a particular attunement to the dynamic nature of software interfaces, or to what Carnegie (2009) calls the interface's 'rhetoric of interactivity' (p. 164). In any case, Lisa clearly constructs herself as someone who sees meaning making as multimodal: language matters, but so do other modes.

Regardless of her multimodal understanding of communication, though, Lisa's position as a UX writer evidently means that her own work is primarily concerned with words and writing; it is therefore in her interest to highlight the importance of the verbal design of software interfaces. And indeed, she goes on to point out that while visual aspects matter, 'the actual words you use are super important' (lines 3). Note the use of the intensifier 'super' to emphasize

[2] Lisa is a pseudonym, as are all the other names I use to describe my interviewees.

the role of verbal content; additionally, the adjective 'actual' here highlights the different, possibly higher status that words apparently (should) have vis-à-vis visual content (see Tognini-Bonelli, 1993, on the use of 'actual'). Ironically, then, although UX writers often claim that the form of language is *not* the point of their work (see Chapter 4), when it comes to making the case for their profession vis-à-vis other semiotic work in software design, the verbal form of their work may suddenly and strategically be emphasized as a distinguishing factor. Making words visible and drawing attention to them as an integral part of software design can thus be another way for UX writers to establish the legitimacy of their work (cf. Chapter 3).

Lisa also positions UX writers as uniquely capable of recognizing the value of words in software interfaces. She repeatedly stresses that her colleagues who work on the (visual) design of software misunderstand the importance of language, saying that 'the design field lost sight of that' (line 4) and that 'they [designers] completely forgot that words are part of the picture' (lines 4–5). Her laughter (line 5) seems to underscore a certain incredulity at how one could forget that communication in/through software interfaces also entails language. In this way, Lisa implies that the professional invisibility of UX writers is not due to a lack of expertise but a lack of comprehension. It is not because of what she does, but because of her colleagues' (mis)understanding of how communication works that her work as a UX writer is valued less than that of designers.

Many of my interviewees emphasized the way language was devalued vis-à-vis other modes. As a second example, I offer the following extract from my interview with Helen. Helen has worked as a UX writer for six years; before this, she worked as a technical writer for two decades.[3] In Extract 5.3, she talks about how her colleagues in visual design (mis)understand the role of language.

**Extract 5.3: (In)visibility in language biography interview (Helen)**

1. One thing I usually talk about when I have an introductory meeting with a
2. designer or anybody I'm trying to show the value of UX writing is that writing is
3. the other half of the design coin. . . . They [writing and design] are of equal weight.
4. And a lot of times it feels like design is elevated and writing is the afterthought,
5. and it can't be that way. . . . [They] work together as two sides of that same coin.

Much of what Lisa does in Extract 5.2 can also be seen here. Helen's main concern is, again, how the skills that are specific to UX writing are undervalued vis-à-vis the semiotic work of her colleagues working on the design of the software: 'it feels like design is elevated and writing is the afterthought'

3 As a reminder, the work of technical writers is similar to that of UX writers, but technical writers typically write texts that are more technical in nature, such as software documentation (see Chapter 2).

(line 4), she says. Note also the similarity in how Lisa and Helen talk about this. Both invoke forgetfulness and a general inattention to words/writing on the part of designers; against this foil, UX writers can then be framed as the ones that are able to understand (and explain) the value of words in software interfaces. Like Lisa, Helen positions herself as capable of recognizing the value of both writing and design. She claims that they are not only equally important ('of equal weight', line 3), but that they are also inseparable: they are 'two sides of that same coin' (line 5).

In this sense, her understanding of communication is not just multimodal but transmodal (cf. Thurlow & Jaworski, 2017b; see also Kress, 1997 on *synaesthesia*) with individual modes such as written and visual communication mutually influencing or conditioning one another. Again, this interrelatedness of (visual) design and writing – and the multi-/transmodal understanding of communication that this is based on – is here framed as something UX writers like Helen understand but their colleagues do not.

Overall, Lisa and Helen both construct a difference in UX writers' and designers' understanding of how communication works. UX writers are characterized as people who recognize that communication is holistic and multimodal, while designers are portrayed as misunderstanding specifically the role of language. There is certainly a degree of exaggeration in how they are explaining this; indeed, Lisa herself mitigates her claim somewhat, saying that it is 'like' designers forgot about words (Extract 5.2, line 4), not that they indeed never think about them; similarly, Helen says that it 'feels like' design is valued more highly (Extract 5.3, line 4). Regardless of this, though, what is interesting is that in their accounts to me as the interviewer, these two UX writers both highlight the differing semiotic ideologies of designers and UX writers as a way to emphasize the unique expertise (and value) of UX writers. These UX writers thus at least implicitly challenge the notion that anyone can write software interface texts. Similar to the explication of metalinguistic knowledge (cf. Extract 5.1), this emphasis on UX writers' understanding of language *and* other modes reframes the otherwise taken-for-granted and naturalized language work (cf. Duchêne, 2020) that goes into writing interface texts as a particular skill that UX writers – but not designers – possess. As such, it is a claim to expertise and professionalism (cf. Chapter 3), highlighting UX writers' supposedly unique value.

Beyond drawing attention to the (multi)modality of their work and its concrete *output*, UX writers also put considerable effort into visibilizing the *processes* of their linguistic labour; recall Extract 5.1, where UX writers are advised to regularly offer metalinguistic commentary on their work in order to increase their visibility vis-à-vis their colleagues. Indeed, visibilizing one's work processes in this way was something most of my interviewees saw as an important part of their day-to-day work, stressing that their work involved not only the writing of interface texts but also what they referred to as 'evangelizing': advertising the value of UX writing within the usually quite large company that had hired them as a UX writer. Six of my interviewees explained that they held regular consultation sessions for non-UX writing colleagues, with the two-fold aim of helping others solve language-related questions and of making them aware of UX writing in the first place. In Extract 5.4, Mona talks about her experience of organizing such consultation sessions upon joining a renowned software company as the first UX writer on the staff. Like Lisa (Extract 5.2), Mona has worked as a UX writer for about a decade and can be considered a seasoned UX writer.

**Extract 5.4: Visibility in language biography interview (Mona)**

1. I would just invite anybody to come in with any piece of writing and I would give
2. them feedback on it. Now, it was pretty sneaky, because what they thought they
3. wanted was, "Oh, here, fix the words for me" or spellcheck it. . . . But what I would
4. do was have them walk through the entire process or, you know, the business goal.
5. . . . So, what I was trying to do was . . . show what UX writing and content strategy
6. could provide.

As Mona explains, her consultation sessions (or 'writing labs', as she called them in our interview) are open to 'anybody . . . with any piece of writing' (line 1), which allows her to potentially reach a large number and a wide variety of colleagues through this practice. As such, her consultation sessions constitute a more oblique way of visibilizing UX writing work: Mona herself calls her practice 'sneaky' (line 2), as she explains that her main goal with these sessions is not to produce concrete linguistic output but to rather focus on broader, invisible processes that make up her language work.

Mona portrays the consultations sessions as a way for her to re-orient her colleagues' understanding of what UX writing is supposed to be about: while they may initially ask her to 'fix the words' (line 3) or 'spellcheck' (line 3) something – tasks pertaining to the verbal form of interface texts – Mona herself uses these requests as an opportunity to instead focus on the communicative context ('the entire process', line 4) and the communicative function these texts (should) fulfil, namely advancing a particular 'business goal' (line 4). The concrete linguistic output, in turn, is deprioritized. One might understand this as a matter of semiotic ideology, this time not in terms of the value of different modes, but rather with regards to the nature of linguistic signs in interfaces and the consequences these signs should produce. For Mona, UX writing appears to be less about words as symbolic representation and more about what these words do or 'provide' (line 6), specifically in terms of business or economic value.

The kinds of consultation sessions that Mona discusses were also mentioned by other UX writers I interviewed, including some that were working at larger companies with more UX writers. Consider Extract 5.5, where Bethany, who has been working as a UX writer for five years, responds to my question about what her typical workday looks like. Note that Bethany works at a large, international software company that employs a lot more UX writers than Mona's employer.[4]

**Extract 5.5: Visibility in language biography interview (Bethany)**

1. A lot of our jobs at [company] even still– a main part of our job is to educate
2. others about what we can do, so that we're not just seen as somebody who
3. comes in and pretties up the placeholder text, but, you know, that we have a
4. perspective and we can add value.

Despite the larger size of the company and the fact that Bethany is neither the only UX writer there nor a recently hired one, a large part of her job (still) consists of 'educat[ing] others about what [UX writers] can do' (lines 1–2). Like in Extract 5.4, what is at stake is again what Bethany's colleagues come to understand as UX writing in the first place. Specifically, Bethany is interested in making sure that she is not seen as someone that 'pretties up the placeholder text' (line 3) and simply polishes what somebody else has written. Like Mona, she thus seeks to shift attention from the verbal form of interface texts to broader, contextual issues. Once again, this is connected to showing how UX writers 'can add value' (line 4), presumably also in an economic sense. Bethany and Mona are not the only ones concerned with this; recall Extract 5.1, where UX writers are likewise advised to 'show the *value* of UX writing' (line 1, emphasis added) in order to gain legitimacy. The same principle seems to be at work here, too: in order to popularize UX writing within a company, the work needs to be framed as profitable.

Overall, then, (in)visibility in UX writing is closely linked to matters of economic profit, all of which points to a 'transactional or entrepreneurial frame' (Urciuoli, 2008, p. 224) that UX writers mobilize to make their work visible in a highly neoliberal environment. (See Marwick, 2013 on neoliberalism in the so-called tech industry.) This neoliberal and typically implicit obligation to prove one's value as a profitable worker is of course something that characterizes much language work (e.g. Urciuoli, 2008; Urciuoli & LaDousa, 2013); it is also something that Redish (1995) identified several decades ago as a key issue for technical writers, a language profession that is often seen as adjacent to or even as a precursor of UX writing (see Chapter 2). In short, in order to gain recognition, creditability, and resources, language professionals have to make their linguistic knowledge visible by linking it to matters of commercial or economic value. Yet, much as the status of UX

[4] For confidentiality reasons, I on purpose avoid giving more concrete numbers about either company.

writers hinges on *visibilizing* the (economic) value of their language work; UX writers' claims to linguistic expertise are also centrally characterized by *invisibility*.

As I have noted before, the work of most elite language workers hinges on its backstage nature – on being kept deliberately out of sight for users/consumers (Thurlow, 2020c). Some wordsmiths manage to articulate exactly the invisibility of their work as a skill, which allows them to claim unique value on the linguistic marketplace (e.g. Mapes, 2023, on speechwriters; Droz-dit-Busset, forthcoming, on copywriters). This, I propose, is also the case for UX writers. Yet, unlike speechwriters and copywriters, who claim expertise by erasing their own authorship, UX writers do not obscure their involvement in the production of software interface texts; rather, they claim to render these texts themselves invisible. As I will discuss in the second part of my analysis, the aspiration of creating invisible interface texts is inherently normative. Before turning to these ideological implications of invisibility, though, I want to examine how UX writers deploy this discourse of invisible writing to position themselves as elite language workers. For this, I return to Lisa's case (Extract 5.2). Recall how Lisa was concerned with distinguishing her own expertise from that of her colleagues. In Extract 5.2, she talked about the different semiotic ideologies UX writers and designers hold in terms of which modes should be prioritized in interface design; Extract 5.6 shows that she extends such distinctions and differentiations to how UX writers and designers approach microcopy (i.e. interface text) itself. This might be understood as a difference in terms of their linguistic ideologies, that is, their beliefs about how language, specifically, is supposed to work in software interfaces.

**Extract 5.6: Invisibility in language biography interview (Lisa)**

1. So one of the things that I've noticed is, like, designers . . . will often want it [the
2. microcopy] to sound really fun and exciting. . . . And then I end up being like a very
3. boring person, I'm like, "Well, actually, we just need them to, like, you know, press
4. this button, then go to the next part." Like, "We don't need to have jazz hands here,
5. you know, and make it so, so noticeably written." So, often, the language work, in
6. a way, it's invisible, like, users don't notice that somebody has this job doing it.
7. And if I do a good job, they won't notice the words at all.

Unlike in Extract 5.2, where Lisa talked about being confronted with her colleagues' belief that language does not matter, she here talks about what happens when they *do* pay attention to language. Bearing in mind that Lisa is constructing a generalized exchange here – something that happens 'often' (line 1) rather than a specific interaction she remembers – her narration of this hypothetical interaction still reveals her own reflexive understanding of how language is supposed to work vis-à-vis her perception of her colleagues' linguistic ideologies. In this case, Lisa proposes that her colleagues tend to *over*emphasize language, wanting it to 'sound really fun and exciting' (line 2) and being generally concerned with making the language stand out. This evidently clashes with Lisa's own opinion. She suggests that UX writing should rather be as inconspicuous as possible: 'if I do a good job, they [users] won't notice the words at all' (line 7).

As I discussed in Chapter 4, the belief or professional value that UX writing should be unspectacular or even boring is not uncommon among UX writers. Indeed, Lisa herself also uses the term 'boring' (line 3) to describe how she thinks her colleagues perceive her. Such disavowals of creative or spectacular writing are, I have argued, one way in which UX writers can demarcate their profession from other kinds of language work, all while raising their own profile in the process. In much the same way, Lisa uses the notion of inconspicuous writing to distinguish herself – in this case, not from other language workers but from her (visual) design colleagues. Concretely, Lisa positions herself as an expert by disaligning with the (constructed) position of her colleagues. She likens their suggestions to 'jazz hands' (line 4), implying that such 'noticeably written' (line 5) microcopy is unnecessarily flashy. Like UX writers' claims that their non-creative writing is better than other people's creative writing, this too might be understood as an act of condescension (Bourdieu, 1991) through which Lisa claims the position of someone with more linguistic knowledge and skill than her colleagues.

The position of expertise and superiority is emphasized by Lisa's construction of herself not only as someone who knows better than her colleagues, but as someone who has to educate her colleagues on how to write microcopy. '[W]e just need them [the users] to ... press this button' (lines 3–4) is apparently the kind of corrective advice she gives her colleagues. Note how the use of 'just' (line 3) further emphasizes the condescending stance. And once again, Lisa's instructional goals entail redirecting her colleagues from focusing on the form of interface texts to their function: what matters are not the words as such but that users will click through the interface in a particular, intended way. Like Mona and Bethany (Extracts 5.4 and 5.5), Lisa thus reorients her colleagues' focus from linguistic form to (business) function, and she does so by writing interface texts that are as inconspicuous as possible. Through this, invisibility becomes linked to value and invisible writing becomes something UX writers can mobilize in order to increase their status as language workers.

Before turning to the more normative consequences of UX writers' invisible writing, I want to briefly summarize my main observations about (in)visibility and status/value in the work of UX writers. As I have established in Chapters 2 and 3, UX writing is – at least at the time of writing – a relatively recent professional role in software design. It is therefore not surprising that UX writers find themselves in need of proving their value both on the larger market of language work but also – or perhaps especially – vis-à-vis their own colleagues. Indeed, the work of their colleagues in (visual) design may not be concerned with language per se, but it is certainly also work that is inherently semiotic. What distinguishes UX writers and designers, then, might be primarily a matter of which semiotic mode they prioritize in their work. As a matter of semiotic ideology, UX writers and designers seem to take a different stance on whether visual or verbal communication matters more in interface design. I have proposed that UX writers operationalize this difference in order to highlight their own expertise on language in interface design as (economically) relevant in a neoliberal marketplace (cf. Urciuoli & LaDousa, 2013). In other words, they render their invisible language work visible by highlighting its business value; and they do so by contrasting their approach to microcopy with the (supposedly) less complex and less nuanced approach of their colleagues. For this, they conceptualize communication as inherently multi- or even trans-modal (cf. Thurlow & Jaworski, 2017b; see also Chapters 2 and 3), an approach that is once again closely aligned with recent theorizations in sociocultural linguistics, and one that strategically decentres language.

Indeed, instead of constructing the value of their linguistic work around the extraordinariness of language, UX writers claim the opposite as their unique skill: the ability to make language so banal and inconspicuous that it becomes invisible. In this way, they gain their status as wordsmiths through their ability to make exactly the invisibility of their language work valuable (cf. Mapes, 2023). At the same time, it is exactly this invisibility that makes the work of UX writers powerful and potentially instrumental in shaping societal norms. In what follows, I thus examine more closely how language is supposed to become invisible; for this, I turn to some of the media ideologies (Gershon, 2010) that structure UX writers' work.

### *5.4.2 Producing Norms: Writing the Invisible Interface*

As I have just shown, UX writers tend to see invisibility or inconspicuousness as a feature of successful UX writing. In this second part of my analysis, I am interested in the normative implications of this ideal and in how, specifically, UX writers claim to make interface texts invisible. This, I suggest, is connected primarily to matters of mediality and materiality. Evidently, it is difficult to conceive of invisibility as a concrete quality of an (interface) text, at least if invisibility is understood as a physical phenomenon (i.e. as something that cannot be perceived by the human eye). But as an *ideal* quality that is ascribed to texts – and often to media more generally (cf. Spitzmüller, 2013) – invisibility is also a discursive phenomenon. Indeed, this discursive ideal has shaped contemporary Western perceptions of media and mediality for a long time; Bolter and Grusin (1999), for instance, trace such discourses of invisibility and transparency back to at least the renaissance. In this vein, I understand invisibility as a central and deep-seated media ideology that structures UX writers' text production.

To illustrate the prevalence of invisibility as a media ideology in UX writing, I again start with a discussion of my blog article data. To recall, these are all instances where UX writers address 'best practices' in their work, I therefore take these articles to be illustrative of UX writers' general or collective understanding of what constitutes good UX writing. In Extracts 5.7–5.10, I offer a compilation of statements by four different UX writers, who all suggest that invisibility is one, if not *the*, hallmark of good UX writing.

**Extracts 5.7–5.10: Invisibility in 'best practice' blog articles**

5.7. When it's crafted well and truly human-centered, your words almost become invisible. People just intuitively get it and know what to do next.
5.8. If you can't remember the text on a button—it's good microcopy. Users shouldn't focus on reading buttons on interfaces; instead, their actions should feel intuitive.
5.9. The best UX writing, meanwhile, is forgettable. ... The language is clear and concise, and it strives to make the experience as frictionless as possible.
5.10. Great writing, like great design, is invisible. What makes things invisible? If it is clear. If it is intuitive. If it is effortless.

Not only do all these UX writers invoke invisibility as a central feature of good UX writing, but they also all offer similar explanations of what exactly constitutes such (supposedly) invisible writing. It is, so these UX writers suggest, writing that requires no (cognitive) effort from the user. We're told that '[p]eople just intuitively get it' (Extract 5.7), that '[u]sers shouldn't focus on reading' (Extract 5.8), that the writing itself should become 'forgettable' (Extract 5.9), and that great (i.e. invisible) UX writing is simply 'effortless' (Extract 5.10). Similar to this emphasis on effortlessness, all but one of the extracts contain the words 'intuitive' (Extracts 5.8 and 5.10) or 'intuitively' (Extract 5.7). What this reveals is that invisibility is typically mentioned not on its own but as part of a cluster of words. Next to 'intuitive' and 'effortless', this also includes 'human-centered' (Extract 5.7), 'clear' and 'concise' (Extracts 5.9 and 5.10), as well as 'frictionless' (Extract 5.9). Together, these examples nicely illustrate how invisibility thus becomes 'mutually enregistered' (Urciuoli, 2008, p. 212, after Silverstein, 2003) with other terms and concepts. By repeatedly being used in the same context, these words come to form what Urciouli calls 'a loose associational chain' (p. 212), whereby one specific term always presupposes the meanings connoted by the other terms. As such, the meaning of 'invisibility' in UX writing expands as it clusters around these other terms, always also evoking the associated meanings of intuitiveness, effortlessness, clarity, and so on.

Particularly telling here is the fact that 'human-centered' (Extract 5.7) is part of the associational chain UX writers establish around the word 'invisible'.

(In other instances in my blog article data, I also found the terms 'user-centered' and 'user-friendly' being used.) The use of the words 'user' and 'human' reveals that these media ideologies entail not only preconceived notions about the software interfaces but also about the users of these interfaces, who should not actively 'read[]' (Extract 5.8) microcopy but simply 'get' (Extract 5.7) what UX writers meant to convey. Once again, this understanding of how texts are read or received is quite interesting, as it is based on a more-than-representational understanding of language and communication (cf. Lorimer, 2005; Thrift, 2007;), whereby the software interface does not need to be cognitively processed by users but is instead processed in an affective and embodied manner. The crux of the matter, however, is that within this more-than-representational understanding of the interface, invisibility appears to be associated with a particular, preconceived role users (should) take on in their interpretation of interfaces or interface texts: that of a cognitively passive but emotionally *affected* recipient whose (re)actions to the interface somehow 'feel intuitive' (Extract 5.8). I will return to this aspect later; for now, suffice it to say that users are typically cast as de-agentialized in these conceptualizations of invisibility.

The mutually enregistered meanings that cluster around the word *invisibility* in UX writing circulate not just in mediatized texts such as these blog articles, they also surfaced in the interviews I conducted with UX writers. Here, too, my interviewees frequently talked about invisibility as a characteristic of good UX writing. What becomes particularly apparent here is how this associational chain (cf. Urciuoli, 2008) makes it possible for UX writers to invoke the ideal of invisibility without explicitly mentioning it. Consider Extract 5.11, where Rachel, who has been working as a UX writer at a well-known tech company for four years, responds to my question about what she would consider good or 'best practice' examples of UX writing.

**Extract 5.11: Invisibility in language biography interview (Rachel)**

1. So I think good UX writing, it's writing that's very, very clear. So, I don't have to
2. think about it as I'm using a product or going through a site or an experience. . . . I
3. don't need to pause.

At no point does Rachel use the words *invisible* or *invisibility* in her response; despite this, she clearly invokes the same discourse of the invisible and intuitive interface that can be seen in the blog article data (Extracts 5.7–5.10). She says that good UX writing is 'very, very clear' (line 1); she equally emphasizes that it should require no cognitive effort: 'I don't have to think about it [the writing]' (lines 1–2). More implicitly, she also alludes to the concept of *frictionless* by suggesting that '[one doesn't] need to pause' (lines 2–3). This shows nicely that through the mutual enregisterment of invisibility with these other concepts, the invisible interface can be invoked without being explicitly mentioned.

Once again, the ideal of an invisible or transparent medium is not unique to UX writing (recall Bolter & Grusin, 1999). By the same token, the ideology of an invisible and intuitive interface is also more widespread and more long-standing. In fact, with regards to software design, the ideal of the invisible interface can be traced back to early scholarship in HCI and to early industry practice in UX design (Pucheu, 2016; see also Petrick, 2020). Here, the invisible interface emerges as a kind of holy grail – something that has structured discussions of (ideal) interfaces for decades (e.g. Norman, 1990; Weiser, 1994; Fishkin et al., 1999) and that is still invoked today (e.g. Patterson & Costain, 2015; Hounshell & Todd, 2016; Susser, 2019). As said, this ideal permeates both scholarly discussions and those of industry practitioners.

In light of these established (or enregistered) discourses of invisibility, Rachel's use of the phrase 'I don't need to think about it' (lines 1–2) takes on additional meaning, as it echoes the title of a famous industry publication, *Don't make me think* (Krug, 2000/2014). Authored in 2000 by UX professional Steve Krug, this book has been widely received in the UX industry and is still today often seen as an important introduction to UX design.[5] More importantly, though, it is one of the key industry publications that has perpetuated the idea and ideal of the invisible interface in UX design (Kendrick, 2005). To be clear, I am not suggesting that Rachel is explicitly referring to Krug's book; my point is simply that given the relationship between UX writing and the broader field of UX design (see Chapter 2), this, too, is part of the associational chain that is at work here: invoking invisibility evokes not only associated meanings of clarity, effortlessness, or frictionlessness but of course also previous, older utterances that have shaped meanings of invisibility in UX design. In other words, the (media) ideological assumptions that shape UX writers' understandings of what constitutes successful examples of their work are part of a more far-reaching discursive formation (Foucault, 1972). It is this that makes the idea(l) of the invisible interface so pervasive and difficult to contest.

In order to illustrate how prevalent the ideology of the invisible interface is, I turn to examining how this discursive ideal surfaces more implicitly in UX writers' discussions of their work. Specifically, I see this as linked to an invisibilization also of the *materiality* of software, which becomes apparent when one examines how UX writers describe the general context and outcome

[5] Currently in its 3rd edition, the book has – according to the author's website – sold more than 600,000 copies and 'has become most people's introduction to User Experience'. Source: *Don't make me think, revisited.* Retrieved from https://sensible.com/dont-make-me-think/ (accessed 30 June 2022). While I don't have any empirical evidence of the book's use and circulation, I can attest that during my own experience working in UX from 2012 to 2021, my colleagues also mentioned this book as a key introduction to UX principles; it is also mentioned in one of the articles in my blog article dataset.

of their work. In my language biography interviews, there is a striking absence of terms such as *technology*, *software*, or even *app* and *website* – words I myself have found central for explaining what kind of work UX writers do. This is not to suggest that UX writers never use these terms, but it is interesting how persistently the UX writers I interviewed described the 'thing' they were working on differently, namely as a *product* or an *experience*. Recall, for instance, how Rachel (Extract 5.11) talks about how one encounters UX writing: 'I don't have to think about it as I'm using a *product* or going through a *site* or an *experience*' (lines 1–2, emphasis added). As said, I am not claiming that UX writers never use technically oriented vocabulary when describing their work; Rachel also uses the expression 'site' (i.e. website) here. Yet, the seemingly disproportionate use of *product* and *experience* across my interview data is telling.

Without making more sophisticated quantitative claims, I offer the following indicative numbers of how my interviewees typically talked about their work.[6] Of the twenty-one UX writers I interviewed, twelve referred to what they were creating as a *product*, eleven of them (also) used the term *experience*, while two interviewees combined these terms and called the object of their work a *product experience*. Not surprisingly, *interface* was a popular term as well – although it was in fact only used by half of my interviewees (ten, to be precise). In contrast to this, the term *website* (including the short form *site*) was used only by five interviewees, the term *app* by three of them, and only one person used the term *technology* and one the term *software* when referring to the general context and objects of their work. The term *(digital) media* that so many digital discourse scholars (myself included) like to use is – perhaps not surprisingly – absent across all my interviews. Finally, it is worth noting that a simple text search of my interview transcripts reveals that the term *computer* occurs only in three interviews, the term *(smart)phone* only in two, and the term *tablet* in none.

As said, I do not want to make any quantitative claims here; my goal is simply to give a sense of how UX writers (in contrast to, say, academics) tend to talk about so-called digital media. What emerges is that terms that connote technological materiality and mediality (like *app* or *software*) are relatively absent in favour of terms that emphasize more ephemeral, non-representational aspects (as *experience* does) or expressions that point to the neoliberal and economic

[6] I compiled this list by manually annotating my interview data. For this, I marked each instance where my interviewees talked about the general outcome of their work and subsequently noted the specific term(s) they had used in each instance. Evidently, individual interviewees often used more than one term for this; therefore, these numbers are not exclusive. I did not annotate the blog article data in the same way, as these articles typically included less (or no) direct statements about what UX writers considered the outcome of their work – unlike the interviews, where I explicitly asked people to talk about this.

context of UX writers' work (as *product* does). *Interface* is of course somewhat of an outlier here, as it might be seen as at least indexing the digital nature of UX writing – though interface itself, too, is a term that obscures the materiality of technology, describing only ever its surface or boundary.[7] Again, there is a link here to the broader field of software design and computer science, where the materiality and physical manifestation of digital media are generally abstracted away (see Borning et al., 2020). Both *product* and *experience* are certainly also indexical of the larger professional field in which UX writers situate themselves. In this regard, *product* is a term that points to the highly entrepreneurial context of software design (Marwick, 2013), while the term *experience* indexes UX writers' orientation to the field of user experience design (cf. Chapters 2 and 3). What both *product* and *experience* obscure, however, is the digitally and technologically mediated nature of UX writing (and UX or software design in general) and, thus, its mediality and also materiality.

In practice, of course, the materiality of digital media does matter. Indeed, sometimes, materiality may be what matters most, such as when people use a smartphone in public spaces (or even just pretend to use one) in order to signal to others that they are not available for conversation (see Ayaß, 2014 on media as involvement shields). At other times, it is the specific channel that matters, such as when making a phone call is deemed more appropriate than writing a text message or vice versa. This, of course, is at the core of Gershon's (2010) media ideologies; the very materiality of media is what makes them semiotically meaningful. However, such meaning potentials that are linked to the material nature of digital media and their mediated-ness are entirely erased in UX writers' discussions of their work. UX writers appear to take invisibility as such a fundamental given of how digital media (ought to) work that the technological medium is taken for granted to the degree that it can be left entirely unspoken and become subsumed under the vague notion of *product* or *experience*.[8]

What emerges as a fundamental media ideology in UX writers' work, then, is their understanding of the materiality and mediality of digital media as merely technical components that can be ignored rather than semiotic resources that contribute to the production of meaning. Thus, the invisible interface becomes a mere carrier or container. Because as Spitzmüller (2015) points out, for a 'thing' to become a 'sign' – or, more broadly, a semiotic

[7] For a more thorough discussion of the immateriality of the interface, see Galloway (2012) and Hookway (2014).

[8] I am aware that *product* might be understood as indexing the materiality of language work by pointing to its concrete (material) production. However, the fact that some of my interviewees explicitly spoke of a *product experience*, which, again, emphasizes immateriality, leads me to argue that product is here primarily a catch-all term that connotes an orientation to entrepreneurialism and value-creation rather than something that indexes the concrete, material production and use of software or software interfaces.

resource – it needs to be not only perceivable but also interpretable. It is in this way that meaning making is centrally connected to materiality; it is through their material nature that signs become legible to social actors (Spitzmüller, 2015).[9] This material condition is exactly what UX writers erase in calling the object of their work a *product* or an *experience*. Overall, this leads to a form of medial invisibility (cf. Spitzmüller, 2013) whereby the software interface is reduced to being a mere and neutral carrier of information. The software interface and the microcopy that forms part of it are neither meant to be perceived nor meant to have any impact on meaning making. What is likewise erased through this, though, is the role of users as social actors who might interpret the interface. To discuss this last point, I now return to examining how users are conceptualized in the ideology of the invisible interface. It is here that one sees most clearly that the invisibility UX writers strive for is ultimately normative, dependent on them shaping – or even prescribing – users' actions.

Importantly, the interface becomes invisible not because users fully understand what is happening; rather, the opposite is the case: the invisible interface depends on users *not* engaging with the underlying technological, social, and semiotic processes. This understanding of users as passive recipients is something I have commented on earlier in this chapter (see Extracts 5.7–5.10); to make this point a little bit stronger, I offer Extract 5.12 from my interview with Sonja. At the time of the interview, Sonja had just started her first full-time position as a UX writer. Prior to this, she had completed a UX writing course and taken several smaller UX writing freelance jobs. Sonja's account encapsulates a certain optimism about the role of UX writers vis-à-vis users, which might be linked to her relatively recent start in this professional domain. Having said which, it is exactly her status as a junior UX writer that makes Sonja an interesting case. As someone who has learnt about UX writing only recently, Sonja is a prime example of how the long-standing discourse of the invisible interface is not just something older generations of UX professionals draw on; on the contrary, it is also very much (re)produced by those that have entered the professional field only recently. These are professional wisdoms or mythologies that are being taught onwards.

**Extract 5.12: Invisibility in language biography interview (Sonja)**

1. You don't want to be seen in the product. You want your writing to be subtle
2. enough that people absorb it without knowing that they're absorbing it. You
3. know, they're using it without really understanding that they– that they've been
4. directed somewhere and or that they've been given an emotional prompt even.

[9] It is here that one sees the close connection between media and semiotic ideologies particularly well; again, my separation of the two is primarily for analytic convenience.

Like many other UX writers, Sonja also strives to invisibilize her writing and/or personal style: '[y]ou don't want to be seen in the product' (line 1), she says. Similarly, she links this invisibility to increased ease of use and reduced cognitive effort for users. A good UX writer, she suggests, writes in such a 'subtle' (line 1) way that users 'absorb' (line 2) their writing without consciously being aware of it. What is particularly evident here is the passive role that is attributed to users. Sonja goes on to underscore this passivity of users by suggesting that they are using software 'without really understanding ... that they've been directed somewhere' (lines 3–4). What emerges is a potentially patronizing stance towards users that is reminiscent of what Massanari (2010) dubs the 'users as victims of bad design' trope, which she sees as pervading much contemporary UX design practices operating under a so-called user-centred design paradigm (p. 404). Here, users are typically characterized as incapable of understanding what software does, while designers are positioned as their 'savior' (Massanari, 2010, pp. 404–406). This is also the case in Sonja's account, where users are only granted a position of affective re-action: they are passive recipients responding to the 'emotional prompt[s]' (line 4) of UX writers without actually comprehending what is going on. They are attributed neither the capacity to comprehend nor the agency to influence what is going on.

One particular belief that is connected to the invisible interface and that underpins Sonja's conceptualization of software interfaces (and digital media more generally) is an understanding of users as passive recipients rather than active contributors to interactions. The asymmetrical and asynchronous nature of the interactions between producers and users of digital media certainly facilitates this understanding. Nonetheless, as scholars in sociocultural linguistics have demonstrated well, audiences are never just passive recipients; rather, they actively shape interaction (see, e.g., the classic work of Goffman, 1981, and Bell, 1984). I will discuss this matter in more depth in Chapter 6, where I suggest that UX writers ultimately engage in *crafting* particular audiences; for now, however, my point is simply that for all their claims to user-centredness, UX writers often seem to entertain a relatively simplistic understanding of the (interactional) role users may take on when using software. In this vein, the user as an active interpreter of the interface is erased; users are instead cast as agentless recipients of the UX writer's directions.

Understanding users as passive recipients, however, obscures the darker, manipulative aspects of what UX writers can and, according to my interviewee Sonja, even *should* do through their work – directing users in such a way that they are not even aware of it. If users are thought of as agentless to begin with, steering their actions can be more easily reconciled as a natural outcome of UX writing rather than explicit manipulation. It is in this way that interface texts are rendered (seemingly) natural and neutral, despite the fact that the decisions

UX writers take are inherently normative, shaping what users can do with and through digital media in the first place. By seeking to create an invisible interface, UX writers thus produce a particular, restrictive interactional situation that leaves little room for reflexivity on behalf of the user and, ultimately, no – or only little – social agency for users.

Evidently, much of the potentially manipulative practices that emerge are due to the particular business context in which UX writers exercise their work. Most likely, it is the companies UX writers work for that are the ones interested in manipulating users (cf. Extract 5.12) rather than the UX writers themselves. As I will suggest in Chapter 6, this may be understood as a form of symbolic violence (Kramsch, 2021; see also Bourdieu & Wacquant, 1992), whereby users are incited to act in particular ways that are supposedly in their best interest while primarily serving the interests of these companies. Amidst this, UX writers, as I have shown in the first part of the chapter, are under constant pressure to prove their (business) value; as such, part of their striving to create invisible interfaces is likely a way of managing this tension. At the same time, though, and as I have shown in the second half of the chapter, this ideal of the invisible interface is also more deeply rooted in the professional practices of the field, including academic scholarship in HCI that influences the profession. It is through the ongoing and often unquestioned circulation of these enregistered discourses of invisibility that UX writers ultimately become complicit in the perpetuation of the ideal of the invisible interface as well as its cultural-normative consequences.

## 5.5 Discussion and Conclusion: (De)Constructing the Invisible Interface

> Interfaces can be consciously designed to create ease for a certain group of people and that "ease" (and power) is predicated precisely on its invisibility for the "average user." (Kendrick, 2005, p. 399)

In her influential work on media ideologies, Gershon (2010) suggests that it is not just users of media who hold particular media ideologies, but that producers equally imbue technologies with particular ideals and intended uses (see also Vidali, 2010; Weidman, 2010). Such assumptions that are embedded in the medium enable and constrain 'how communication can take place through that medium, how the communication circulates, and who can participate' (Gershon, 2010, p. 284). In this chapter, I have focused on one such assumption, namely the ideology of invisibility in UX writing, which impacts both UX writers' status as language workers and has fundamental consequences for how their work impacts users of digital media. As Kendrick (2005, quoted above) reminds us, the 'ease' that comes with invisibility is always tied to particular assumptions about the average – or ideal – user.

My main objective in this chapter has been to document how UX writers understand and negotiate the supposed invisibility of their (elite) language work, and to discuss the extent to which they understand this invisibility as linguistically and culturally consequential. As I have shown, there are two interconnected kinds of invisibility in UX writing. The first has to do with the (in)visibility of their professional role and the way language as a mode is valued in software design. In other words, this is a matter of semiotic ideology (Keane, 2003, 2018), and it is connected to how UX writers have to establish their value and status not just vis-à-vis other language workers but also – or especially – vis-à-vis their colleagues in (visual) design. In this regard, I have discussed how UX writers claim status by making the invisibility of their language work visible and valuable and that they do so by framing the work of their colleagues in visual design as semiotically less complex. Invisible writing is a discursive ideal UX writers operationalize in order to establish the (also economic) value of their work. This, in turn, is linked to and sustains a second kind of invisibility: that of the interface. I have discussed this primarily as a matter of media ideology, arguing that the ideology of the invisible interface is not only persistent in UX writers' work, but that it is also central in determining how ordinary users can (or cannot) use digital media. These two invisibilities are inevitably connected; this shows once more how semiotic and media ideologies are inherently interrelated (cf. Thurlow, 2017) – the discursive ideal of invisibility is what allows UX writers to make particular status claims, but this idealized invisibility is also what makes the work of UX writers instrumental in shaping societal norms.

While invisibility is typically associated with lower-status work (e.g. Duchêne, 2011), UX writers turn invisibility into a status claim. This parallels the way these professionals also turn non-creativity into a marker of prestige (cf. Chapter 4). In this sense, (non-)creativity and (in)visibility are closely connected: Just as it is the control over whether one is creative or not that can turn non-creativity into a marker of prestige, so is it the choice between visibility and invisibility – that UX writers appear to master so expertly – that allows one to claim status based on being invisible. Indeed, as Thurlow and Jaworski (2013) show in their work on elite discourse, it is precisely the balancing act of (in)visibilization – of carefully controlling what is rendered visible and what is rendered invisible – that is a key marker of eliteness. What emerges, then, are the complex (and multiple) hierarchies that govern invisibility and eliteness. UX writers are *relatively* speaking more privileged (for instance, vis-à-vis other lower-status language workers like subtitlers), and it is important to recognize that the invisibility of UX writers (unlike that of, say, subtitlers) is self-chosen – just as the visibility of, say, a celebrity author is self-chosen. This goes beyond the concept of anonymity (i.e. authorial credit) and instead includes also matters of social invisibilization (e.g. Star & Strauss,

1999; see also Thurlow, 2020c). It is in this way that UX writers' invisible words are ultimately implicated in wider, cultural-political issues of power and inequality.

The 'ease' or 'frictionlessness' UX writers connect with invisibility is always tied to an exercise of power and the production of social and cultural norms. It is here that UX writers *misrecognize* (in a Bourdieusian sense) the invisible interface. Misrecognition in this sense is not the same as misunderstanding; rather, it is a social process whereby particular practices and actions are not recognized (i.e. not perceived) for what they are. In this case, the invisible interface is recognized by UX writers as creating ease-of-use for users but not as an act of normalizing particular uses and users; therefore, its culturally normative effects are maintained while (or because) they are being concealed. Importantly, misrecognition (*méconnaisance*) is based on the knowledge (*connaissance*) specific to a social field; it is tied to 'the set of fundamental, prereflexive assumptions that social agents engage in' (Bourdieu & Wacquant, 1992, p. 168; see also Hanks, 2005, pp. 77–78). The misrecognition of the invisible interface is thus not an individual act; it is a social process that is more deeply rooted in the values that circulate in the field of UX writing as a profession.

In terms of professional values, the discursive ideal of the invisible interface is not unique to UX writing, it is a long-standing ideological norm in HCI and UX design, which in turn is linked to even older ideologies of media transparency and immediacy (cf. Bolter & Grusin, 1999). In closing, I want to return to one of the most influential texts that has shaped this ideal in UX design, Norman's (1988/2002) well-known book *The Design of Everyday Things.*[10] As it happens, this text was also mentioned in one of the 'best practice' articles in my blog article data. Tellingly, Norman (1988/2002) suggests that technology becomes invisible when designers craft obvious and natural interfaces or controls. He urges readers to produce what he calls 'natural design' (p. 4): 'there must be a close, *natural* relationship between the control and its function: a *natural mapping*' (p. 28, emphasis in original). Throughout, Norman never questions the ideological nature of this invisibility, nor does he acknowledge that what counts as 'natural' is in itself implicated in socially formed norms – that the so-called natural user interface is in fact not natural but rather *naturalized* (see also McCorkle, 2017). As such, one of the most influential texts in the UX industry treats invisibility as entirely unproblematic

[10] Norman might be the most fervent proponent of the invisible interface. His publications also include the book *The Invisible Computer* (Norman, 1998), whose dust cover blatantly states that 'technology should be invisible, hidden from sight'. I have chosen to discuss *The Design of Everyday Things* here, however, as it is, to my knowledge, more popular – certainly among UX professionals.

and apolitical. This point of view seems to persist in UX writing until today, as I have shown in the current chapter.

Early editions of Norman's (1988/2002) book close with his speculations about 'the invisible computer of the future' (p. 185ff.). Musing about a calendar that not only records appointments but proactively informs users of things such as scheduling conflicts, Norman suggests that in the future 'the computer [will be] invisible, hidden beneath the surface; only the task [will be] visible' (p. 186). Ironically, the kind of invisible calendar/computer Norman imagined in the 1980s is not too different from calendar applications that are pre-installed on most personal computers and smartphones nowadays. Contemporary calendar apps allow users to send and receive invitations, alert them to scheduling conflicts, and remind them – automatically – of national holidays and even of the birthdays of people in their phone's contact list. But for all their ease of use, these calendars only appear invisible or feel natural because they conform to so many Western norms – starting with the fact that their default view typically displays a time window between 09:00–17:00, corresponding to the average working hours in the United States and much of Western Europe. This may sound practical to people working in Western countries, but apart from being Eurocentric, it also entails particular unspoken norms about who should be using a calendar (perhaps someone who works an office job) and what they should be using this calendar for (presumably to schedule work appointments). This may seem like a trite example and a harmless matter, but the ramifications are significant. The 'invisible computer' is inherently and unavoidably based on naturalized but no less ideological norms, and the 'user-friendliness' of the interface is ultimately nothing more than a 'constant constriction and interpellation' (Chun, 2011, p. 67). Ultimately, any invisibility is predicated upon its ideological basis and exerted through normative effects. If anything is indeed invisible, it is these effects, which appear to be largely ignored or unseen by producers of digital media even today.

The invisible interface is not new, and it has not gone uncommented on, especially by scholars in critical media studies (e.g. Wysocki & Jasken, 2004; Kendrick, 2005; Carnegie, 2009). In earlier work, rhetoric and media scholars Wysocki and Jasken (2004) specifically examine how interfaces are (not) seen by their producers. In their analysis of handbooks and guides on digital writing from the 2000s, they find that these instructional texts invariably prioritize technical efficiency over social meanings and functions. This in turn leads them to argue that producers of interfaces are taught to think of interfaces as functional and arhetorical. By contrast, my own analysis of UX writers' understandings of their work suggests that they seem to be more aware of the rhetorical force of their work. The language biography interviews I have analyzed in particular reveal that UX writers understand perfectly well that their writing influences and persuades users to act in particular ways; indeed, as I have shown, this is precisely what UX writers seek to do through their writing. What they seem to be less conscious of, however, are the cultural-political implications of their work. The invisibility UX writers strive for evidently hinges on privileging a particular kind of seeing, a particular position from which things are rendered visible (or not). This is a kind of sociopolitical ramification that UX writers appear to be less aware of or troubled by.

The 'frictionless experience' UX writers strive to produce is always enmeshed with the production and re-inscription of particular norms, and the discursive ideal of invisibility that structures UX writing is based on a *misrecognition* of technology as neutral and apolitical. All of this points to a simple but important fact: communication with and through (digital) media is shaped not only by users' perspectives but also by the semiotic and media ideologies of its producers. Importantly, these ideologies have normative consequences, as they are implicated in the way that producers of digital media design particular kinds of (inter)action. In Chapter 6, I turn to investigating the normative impact of UX writers' semiotic and media ideologies more concretely by discussing how they stylize and craft particular audiences through their work, thereby determining who should be using digital media – and who not. Ultimately, this final analytic chapter thus demonstrates that interfaces are normative – or political – not just as texts that can reiterate existing, ideological views but also as interactional encounters whereby the joint actions of technology and human(s) reproduce hegemonic norms.

# 6 Crafting an Audience

## UX Writing, User Stylization, and the Symbolic Violence of Little Texts

As I have shown in Chapter 5, the beliefs that UX writers hold about how language works – or ought to work – impact not just the practices and processes of their work but also the textual products that they create, and, ultimately, the recipients of these products, that is, the users of digital media.[1] In this final substantive chapter, I investigate the normative impact of UX writers' semiotic ideologies and their ramifications for users in more depth. As such, this chapter returns to the linguistic products that UX writers create by discussing texts in digital media interfaces (so-called microcopy, see Chapter 2). My analysis focuses on a particularly impactful example of this work, the microcopy produced for cookie consent notices. While certainly not representative of all UX writing, I believe that this kind – or genre – of text is a particularly productive genre to examine as it is one that aptly shows the significant ramifications that UX writing can have for users. Concretely, I am interested in how these 'little texts' (Pappert & Roth, 2021) in software interfaces construct certain kinds of audiences and participant roles, which leads to particular structures of interactional as well as symbolic in- and exclusion. Examining the kinds of audiences and addressees that surface in and through these texts, I suggest, can help scholars consider how digital media entail not just traditional notions of audience design (cf. Bell, 1984) but also a more explicit and active crafting of audiences done by the producers of software, whereby some people are constructed as audiences and others not.

In what follows, I address these different, crafted audiences as they surface in the work of UX writers. I examine these aspects of audience crafting by attending to participation structures, user stylization, and designed affordances. The chapter is structured as follows. I first give an overview of each of these concepts and how they are implicated in audience crafting. I then offer an analysis of the microcopy that UX writers produce for cookie consent notices to illustrate how audience crafting takes place. Specifically, I discuss how automated participant roles, the stylization of users, as well as the design of

[1] Discourse, Context & Media (Portmann, 2022).

imposed interaction lead to an encoding of both specific participant roles as well as particular social identities in software interfaces. I suggest that this may be understood as a form of symbolic violence (cf. Kramsch, 2021), whereby the software interface is used as a means to impose not just an interaction order but also a particular social order onto users. Throughout my analysis, I connect this case study to my previous discussion of UX writers' semiotic and media ideologies (Chapters 4 and 5), showing that their own understanding of how language works is enmeshed with the re-inscription of particular, hegemonic norms. It is by positioning users in particular ways and in particular moments that interface texts exercise their power and that these little texts are, ultimately, much bigger than one might imagine.

## 6.1 Participation and Audience in Digital Contexts

Digital media have given rise to a range of new social practices and ways for people to interact with one another. This has led scholars early on to ask whether digital media create new kinds of audiences (cf. Livingstone, 1999). Certainly, these new practices challenge how scholars of language are thinking about questions of audience and participation. Most sociolinguistic scholarship in this regard is based on the classic work of Goffman (1981) and Bell (1984), who above all show that audiences are not passive recipients but active contributors to interactions. Technology, however, changes interaction, and with this, participation structures.

The notion of audience, specifically, has been complicated in light of digital and, in particular, social media, where 'context collapse' (Marwick & boyd, 2011) gives rise to new kinds of networked audiences. In this vein, Androutsopoulos (2014) examines audience design on Facebook, where users are confronted with the overlapping linguistic repertoires of their 'collapsed' audience. Drawing on Bell (1984), he shows that audience design on Facebook is not entirely different from audience design in offline contexts, but it is complicated, with context collapse leading to new kinds of audience configurations. Similarly, Boyd (2014), in discussing participation on YouTube, highlights that participant roles may take on new or different meaning in digital environments. Specifically, he suggests that 'ratified' takes on new meanings when nuances such as whether someone is registered or not add dimensions that offline, co-present interaction does not entail. With regards to production roles, Draucker (2015) discusses participation structures in Twitter interaction, arguing that in addition to Goffman's (1981) roles of animator, author, and principle, Twitter interaction can also include a *broadcaster* role, someone 'responsible for distributing the talk to others' (p. 51) without being responsible for the actual production of the talk (as Goffman's animator is). As Draucker points out, such a distinction may not be needed for examining co-present, spoken interaction, but is useful for understanding interaction on Twitter. This limited overview of how Goffman's and Bell's work has been applied to digitally mediated interaction is far from exhaustive, but it illustrates two points. First, it shows that while existing understandings of participation and audience do not translate directly to digitally mediated environments, they still serve as a useful starting point for examining how and what kinds of production and reception roles are made relevant. Second, this work illustrates that the medium itself does not determine but nonetheles impact what kinds of participation structures can arise.

In digital contexts, questions of audience and participation have further been complicated by the role that software itself starts to take as a non-human participant. In this regard, Eisenlauer (2014) discusses how Facebook interaction may include texts that are created in a fully automated fashion, such as time stamps that get added to status updates. Thus, Facebook becomes what Eisenlauer calls a 'third author' – here, software takes over the production of the text. Similarly, algorithmic participants can also take on an audience role. This way, Jones (2021) points out, digital media contexts change the kinds of audiences that ordinary participants have to account for: our utterances on social media sites are also 'listened to' by algorithms and, through them, by corporate actors. Again, these kinds of non-human actors behave differently than the kinds of participants that Bell and Goffman envisage in their models, which can lead to new kinds of participation structures with new kinds of consequences. This is what Jones (2020b, 2021) addresses with his proposal of algorithmic pragmatics, which concerns 'the pragmatics of human-algorithm communication' (Jones, 2020b, p. 19) and hence acknowledges that meaning and action in digital communication are increasingly influenced not just by the communicative contexts that digital media provide but also by the software itself as a non-human participant.

While Jones (2020b) is interested in the role of algorithms, my concern here lies with a different aspect of software, namely the front-end interface and its production. Specifically, my point is that it is not only users that are implicated in participation structures but also the producers of software. As I have discussed in previous chapters, in digital discourse studies, the role of professionals such as UX writers, designers, or software engineers – that is, the people producing digital media – is still discussed relatively rarely. They are frequently treated as an unknown or abstract entity, despite the fact that they significantly impact what kind of participation can (or cannot) take place. In this sense, and as Anderson and Borges-Rey (2019) illustrate, audiences are also constructed in the design process of digital media, and specific ways of engaging with digital media are thus always already present in the interface itself. In short, people's possibilities for performing certain actions and certain identities are shaped by producers and users. However, producers' imagined audiences can be quite problematic, as they are often motivated as much by internal politics as they are by the realities of actual users (Massanari, 2010). What is more, these audiences that are expressed through the interface have normative consequences as they reproduce a particular, typically white and Western, worldview (see also Chapters 1 and 5), which is why I suggest that they deserve more attention from digital discourse scholars, specifically with regards to how users are thus stylized as particular addressees.

## 6.2 User Stylization

Digital media interfaces always hail certain people as users and others not (cf. Althusser, 1971). It is in this way that interfaces also produce particular audiences by constructing an 'ideal' addressee or superaddressee (Bakhtin, 1986; Piller, 2001). This, I suggest, is connected to how UX writers engage in the *stylization* of users. In sociocultural linguistics, stylization is closely linked to the concept of style and styling. Traditionally, *style* was understood as referring to speakers' linguistic variation according to specific speech situations (e.g. Labov, 1972). Today, style is typically understood as more dynamic, not as a passive attribute of speakers but as a resource that people can use to (actively) style themselves (cf. Eckert, 1996); it is also frequently understood as something that is accomplished multimodally (e.g. van Leeuwen, 2005). In other words, style is seen as something that people do through a variety of modes in order to position themselves as certain types of social actors. Style – or *styling* – is thus a resource for identity construction (e.g. Eckert, 1996; Cameron, 2000b; Coupland, 2001). As Bell (1984) proposes, however, style is also linked to issues of audience in that stylistic choices are centrally motivated by who speakers are addressing. In this sense, style as audience design is a dynamic and implicit phenomenon, with speakers constantly and often unconsciously adapting their style for and in response to a particular audience. However, as Bell notes, speakers can also initiate a marked style-shift which involves 'address[ing] persons as if they were someone else' (p. 186). It is in this way that audience design may also involve a more deliberate *stylization*.

While styling is typically understood as something that is part of ordinary, everyday interaction, Coupland (2001) takes *stylization* to be a more explicit, strategic act of projecting a persona via the use of recognizable sociocultural styles. A key feature of stylization is hence the act of 'putting on' someone else's voice (Coupland, 2001). Other scholars, however, use the term stylization for slightly different, more specific purposes. Cameron (2000b) and Thurlow and Jaworski (2006), for example, treat stylization as something that is done to people by others. In this sense, stylization is a top-down, prescriptive practice (Cameron, 2000b) that, as Thurlow and Jaworski (2006) put it, entails the 'the imposition of a style' (p. 194) onto others. It is this more deliberate, explicit use of stylization that I adopt in the current chapter, understanding it as the strategic fabrication and projection of a recognizable sociocultural style by one agent onto another. Specifically, I understand UX writers as stylizing users by imposing a particular 'built-in' social identity that also establishes (or enforces) a particular interactional role for users and thus a predetermined or at least constrained participation structure.

UX writers thus do not only implicitly write for an audience but also actively invent and craft their audience, producing what Bakhtin (1986) calls a *superaddressee*. Bakhtin (1986) suggests that utterances are always dialogic and directed *at* someone; as such, they entail the anticipation of a listener and, being shaped by this anticipation, include the addressee as such. Importantly, Bakhtin understands utterances not only as oriented towards an actual, real addressee but also as oriented towards a 'superaddressee': a projected, ideal listener that is imagined to fully and perfectly understand an utterance (Morson, 2006). In this way, Piller (2001) sees Bakhtin's work as particularly illuminating for understanding the ideal subject positions that texts construct: any text, software interfaces included, constructs the position of the ideal recipient. Essentially, this is the logical complement to Bell's (1984) audience design: speakers do not only adapt their utterances to their audience, the audience itself is also configured by the speaker. Through the utterance or text, some people are made into audiences, while others are not. In other words, interfaces always entail a superaddressee, a subject position that is created through user stylization and that users are asked to take on.

## 6.3 Affordances and Designed Interaction

Digital media interfaces impose not only social identities, but through their (designed) affordances, they also impact users' actions. It is in this way that UX writers, together with their colleagues in adjacent professions, quite literally shape what users do with and through digital media (cf. Moschini & Sindoni, 2021, who make similar observations about professionals in software development). In short: interfaces have affordances that impact our actions. My understanding of affordance is rooted in the classic work of Gibson (1979), who coined the term to describe how objects frame – but do not determine – what organisms can do. While Gibson concerns himself primarily with the affordances of natural objects, affordances can also be 'designed into the artefact' (Hutchby, 2001, p. 449). Drawing on Gibson, Hutchby emphasizes that affordances are both functional and relational: functional in the sense that they enable and constrain certain actions and relational in the sense that the affordances of an object differ for different organisms. Affordances are thus actualized in relationship with the individual organism while existing prior to and independent of the organism (cf. Hutchby, 2001, p. 448).

While Hutchby (2001) is best known for bringing the term affordances to studies of language and communication, the concept also has a longstanding tradition in human–computer interaction (HCI), where Norman (1988/2002) is known for applying affordances to the design of tools and objects. Taking a more use-oriented approach, Norman argues that things ought to be designed in such a way that their intended use is obvious, and for this, designers should pay attention to affordances in the design process. As such, Norman's view of affordance diverges from Gibson's conceptualization; Norman is primarily concerned with making the possible uses of objects visible and instrumentalizes the concept (see Scarlett & Zeilinger, 2019; for a comprehensive critique of Norman's approach to affordance, see also Lialina, 2019). Again, though, affordances are relational and are thus determined by the designed object *and* the user. Hyperlinks, for instance, obtain their affordance as links, on the one hand, because web browsers have been coded and designed to identify and display hyperlinks but, on the other hand, also because people perceive them as hyperlinks – that is, as something that can be clicked – in the first place. (Hopkins, 2020, p. 4). Consequently, the concept of affordance allows scholars to understand interaction as determined by both producer and user.

Recently, scholars in media studies have suggested that digital media complicate traditional notions of affordance due to their dynamic and malleable nature (Bucher & Helmond, 2018) and their 'different-yet-intersecting layers of materiality' (Scarlett & Zeilinger, 2019, p. 18), entailing both the physical layer of computer hardware and the virtual layer of the software interface. Kirschenbaum (2012) thus speaks of 'forensic materialism', which includes anything located in the physical world, and 'formal materialism', which pertains to simulated materiality via, for instance, software processes. The affordances arising from the simulated materiality of front-end interfaces of software make the ideological underpinnings of designed affordances particularly explicit: these are affordances whereby 'technically *possible* uses become more or less *normative* through productive constraint' (Stanfill, 2015, p. 1062, emphasis in original). Taking affordance as a lens, it is thus possible to examine how interfaces establish what users *should* do – and how producers of software like UX writers encode such interactions in the software interface itself. This brings me to my case study, where I discuss how particular audiences and interactions become encoded in and through UX writing.

## 6.4 The Data

For this case study, I turn – finally – to some of the linguistic products that UX writers create as part of their work, looking at some of the microcopy typically produced by UX writers. To be clear, and as I have discussed in depth in earlier chapters of this book (see especially Chapter 2), most UX writers see their work as more complex than this; their work often also includes conceptual tasks pertaining to information architecture and content governance, which may result in non-user facing texts such as content models, style guides, or copy docs. Nonetheless, microcopy is the most emblematic and a rather consequential linguistic product that UX writers create, as it directly affects the kinds of interactions that become possible for users.

My analytical focus lies on one particularly impactful example of microcopy: the microcopy produced for cookie consent notices. Deceptively small or fleeting, these texts are a typical example of a 'little text' (Pappert & Roth, 2021; cf. also Halliday, 1985): small in size, highly contextualized, and tied to a specific purpose (Hausendorf, 2009). While perhaps not an obvious site for exploring audiences and interaction (after all, there is no interaction with other human participants in cookie consent notices), my point is that cookie consent notices as 'little texts' also construct different kinds of participants and make different kinds of participant roles and subject positions available when people interact with and through them.

As a genre, cookie consent notices are certainly not representative of all UX writing. Still, I find the microcopy of these little texts particularly productive for scholarly analysis, not least because cookie consent notices have significant ramifications for users: what is at stake is nothing less than their privacy, even though these texts often tend to be phrased as banal and insignificant encounters that users shouldn't give much thought to. Indeed, as I will show in my analysis, it is here that UX writers' strategic fore- and backgrounding of specific interactions and the thus ensuing consequences for users that I discussed in Chapter 5 become especially apparent. Beyond this, my reasons for analyzing cookie consent notices are also in part practical: unlike, say, sign-up flows or error messages that require extensive engagement with software, cookie consent notices can be easily collected and compared across a variety of websites. Additionally, it is a genre of microcopy most readers (especially those living in Europe) will be familiar with. I also want to reiterate at this point that not all cookie consent notices users encounter are written by a UX writer. As I discussed in Chapter 5, the linguistic work UX writers produce is often seen as naturalized (i.e. innate rather than learnt); as such, other people that are involved in the production of software interfaces may end up writing microcopy as well. Still, my point is that the practice of writing microcopy, including the microcopy of cookie consent notices, is part of the professional domain of UX writing and typically also understood as such by UX writers themselves.[2]

[2] Despite the fact that cookie consent notices may also be outsourced to Consent Management Platforms, which provide code snippets that website owners can buy and integrate in the codebase of their website, my fieldwork suggests that UX writers still understand the microcopy in cookie consent notices as part and parcel of their own professional domain.

In the analysis that follows, I draw on a convenience sample of 151 English-language cookie consent notices, collected between October 2020 and November 2021. These cookie consent notices reflect my own encounters with such texts across the mentioned time period. While I initially collected all examples that I came across, I eventually shifted to collecting only particularly salient examples, that is, examples that stood out as different from what I had seen before, with the goal of capturing a broad range of examples. As such, my dataset includes cookie consent notices from a wide range of national and international websites, all of which were accessed through a Swiss IP address. My choice to collect only English-language examples is motivated by the fact that many UX writers work only on English-language texts. (Texts in other languages are often translated by localization experts.) To be clear, my data is typical rather than representative (cf. Thurlow, 2006), since my interest is not to provide a comprehensive account of cookie consent notices, but rather to offer a critical perspective on issues of audience and participation in the work of UX writers. In this sense, my aim here is not to showcase all of this data; instead, I discuss a few select examples in depth.

To complement my textual analysis, I once again draw on other data that I collected during my research, specifically, the language biography interviews (see Chapter 4) as well as a handful of second-hand, public interviews with UX writers that I collected during my mapping of the profession (see Chapter 2). I here use this discourse ethnographic data to situate the work of UX writers and to understand the way they think about and understand the role of audiences in their own work. Finally, and to substantiate my claims with regards to how users respond to these texts, I draw on existing empirical studies from the field of HCI that discuss users' reception of interface texts. As said, however, my primary focus in this case study lies not on users but on UX writers as the producers of digital media, and, specifically, on the notions of audience and participation that structure their work. This in mind, I start by considering the kinds of participant roles that emerge when users encounter the product of UX writers' work.

## 6.5 Analysis: Crafting Audience in the Work of UX Writers

### *6.5.1 Automated Participant Roles*

When people interact through digital media, they always interact on two levels, with other humans *and* with the software, both of which influence their actions. Digitally mediated interaction hence always entails a 'pragmatic duality' (Sjöström & Goldkuhl, 2004), which also impacts the kinds of participant roles that surface. I will discuss this by considering how websites

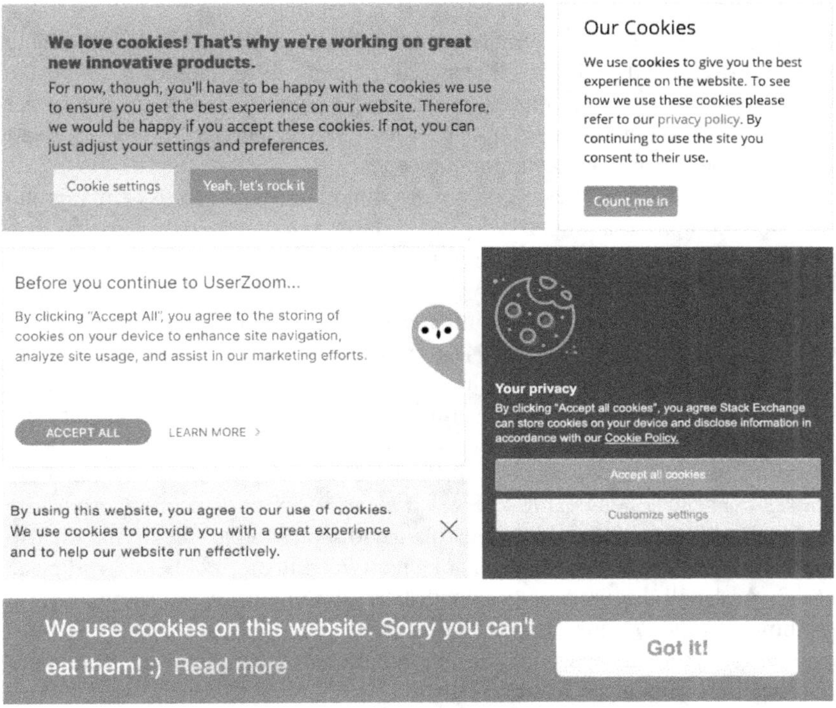


**Figure 6.1** Cookie consent notices. From left to right, top to bottom: Kokojoo/Dayog Kabore; Titan Comics; UserZoom/Usertesting Technologies, Inc; Stack Overflow; Sage Publishing; Boardgame Arena

ask users to accept cookies. To this end, Figure 6.1 offers a compilation of cookie consent notices from my dataset.

A cookie is a small text file that a web server can store on a user's computer. Cookies are used to remember information about the user, for instance, the user's location or the username that was last used to log in on a page. More complex cookies can also be used to record people's browsing behaviour, which can later be used to make inferences about the kinds of audiences that visit a site. However, this is rarely what people consider when they consent to cookies. Indeed, the different cookie consent notices in Figure 6.1 hardly inform users of what cookies are, let alone what kinds of technical interactions (cf. Carmi, 2021) one is asked to consent to. Most of these texts are purposefully vague about the matter, speaking simply of offering 'a great' or even 'the best experience'. Many make use of emphatic language ('We love cookies!'; 'Count me in'; 'Yeah. let's rock it') and playful, tongue-in-cheek references to actual cookies ('Sorry you can't eat them!'), all of which construct a particularly (and, given the legal implications of these texts, somewhat strangely) enthusiastic tone. At the same time, however, and as Ma & Birrell (2022) show in an empirical study, the language used in cookie consent notices has a significant impact on whether people choose to opt-in or opt-out of cookies. Clearly, language is of key interest in these little, seemingly fleeting interface texts.

Intentionally or not, these cookie consent notices thus tend to be manipulative, something that is in part achieved through semiotic means. Beyond this, though, they are of course also highly creative. Indeed, at first glance, these examples seem to contradict UX writers' claims that their work is not creative (cf. Chapter 4). Certainly, many of the examples in Figure 6.1 exhibit the kind of 'literariness' that Carter (2004) sees as typical of (linguistic) creativity through the use of metaphor and humour. But more important than this, these interface texts are also creative in a discursive sense (cf. Jones, 2010): they are clearly meant to *do* something – or, rather to get *users* to do something, namely to accept cookies. In this way, the linguistic and discursive creativity that these cookie consent notices exhibit serves a clear business or economic purpose (again, see Chapter 4 on UX writers' claims to such strategic creativity; cf. also Gormley, 2020, on the co-option of creativity into a neoliberal workplace ethic). As I will discuss in more depth later, the linguistic and semiotic form of these little texts thus obscures what other purposes cookies may serve – especially, what their purpose may be for corporations – and what kinds of ramifications clicking 'Got it!' might have. Microcopy is here used to creatively invisibilize particular (inter)actions towards users. Indeed, on a more basic level, cookie consent notices, usually implemented as modal popups, also create a specific interactional situation, whereby users are positioned in a relatively constrained participant role – one where they often have no other option than clicking agree (or turning away).

The constraint on participation is linked to how interactions like these are entirely determined by *when* such texts appear. There is an issue of temporality at work, whereby users are asked to be obedient in the face of urgency. As Jones (2020a) points out, these popups usually appear just when people wish to engage with what they obscure so that '*not* trying to adjust one's "cookie preferences" [becomes] the rational choice' (p. 94, emphasis in original; Gray et al., 2018, on 'consent walls'). Clicking 'agree' thus constitutes an action that means different things to the user and to the technological system. To the user, it's a quick click made in order to continue to what they actually came for; to the system, it constitutes the permission and the trigger to send specific cookies to the user's hard drive. The computational process and the textual performance diverge. Here, then, we have a different kind of automated action. This is not an automated text action in Eisenlauer's (2014) sense, but rather automation in the sense that people become conditioned to more automatically take the action that the software and those who built it would have them take. In this case, the software interface as well as its underlying technological system (i.e. the algorithm and its actions) significantly shape the users' interaction and, hence, the kinds of participant roles that they may take on in the interaction itself.

### 6.5.2 *Stylizing Users*

Cookie consent notices do not only construct specific participant roles, they also construct particular subject positions for users. In other words, through these texts, users are also stylized in particular ways. Consider the cookie consent notice in Figure 6.2: not only does this interface give me only one kind of action that I can take (agreeing to cookies), it also makes me take this action in a very particular way, that is, by declaring that 'I'm cool with cookies'. Through this, the interface establishes a particular identity that I am supposed to perform in this exchange. This becomes particularly clear when trying to imagine who might utter a statement beginning with 'I'm cool with . . .'. It's much easier to imagine this being said in an informal exchange between two friends than in an encounter between a judge and a lawyer. This is not to say that such an utterance would or could not be made in the latter instance, but it is generally associated with social encounters and social identities of the first type. It indexes a particular register (cf. Agha, 1999), and, hence, a particular speaker, but also a particular addressee. In this case, the website of this company was written in the same casual style, and I assume that this constitutes a deliberate stylistic choice – the company in question is an artificial intelligence start-up that might well want to construct a particular image for itself. In other instances, I found cookie consent notices that employed a more formal style. However, the point that I want to make here is that *any* style choice, even a supposedly 'neutral' one, will index a particular register and hence create a particular subject position. It is in this way that users are stylized into predetermined social roles. Here, then, the impact of UX writers' work on audiences and addressees becomes especially apparent.

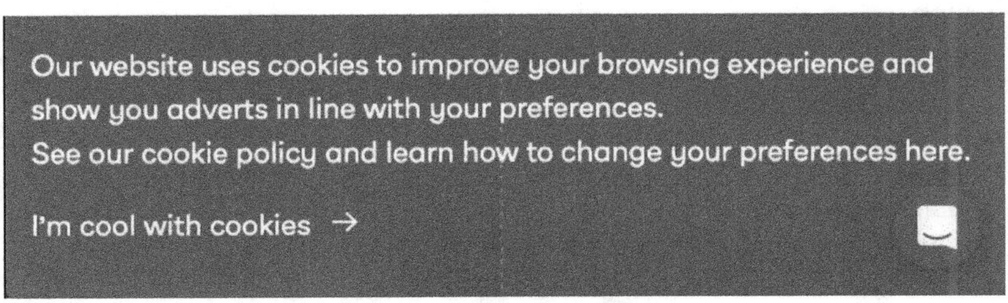


**Figure 6.2** Cookie consent notice: Crystal AI

For UX writers, knowing their users (i.e. their audience) is a key concern. For this, they typically engage in a range of research practices, which, to name but a few, may involve interviewing users, showing them draft copy or doing usability testing (often in collaboration with dedicated UX researchers). In this sense, UX writers write their texts *for* a particular audience, engaging in more or less conscious audience design (Bell, 1984). However, I argue that UX writers not only write for an audience but at the same time also invent and craft said audience through their work. To be clear, it is rare that UX writers are the only ones that impact microcopy (see Chapter 2); their work is often quite collaborative and may involve colleagues from business, design, engineering, or, in the case of legally sensitive content such as cookie consent notices, also legal experts. Yet, while these other professionals may impact the text, it is UX writers that are seen as the author of the final microcopy, and it is they who are responsible for its (linguistic) style – and, thus, for the strategic stylization that is done through microcopy. It is in this way that they construct a superaddressee (Bakhtin, 1986), a social identity that users, if they wish to use that software, have no choice but to take on.

By attending to user stylization, it becomes clear that the audiences that are designed and configured in the production of these texts are also social identities and, hence, a matter of normativity and power. There is a politics to these interfaces (Selfe & Selfe, 1994; see also Djonov & van Leeuwen, 2017), one that is, I suggest, particularly important to address in order to understand how people's interactions with and through software are also shaped by how professionals like UX writers 'design' certain people as users so that some people are made into (legitimate) audience members while others are not. In this way, the normative uses and users are demarcated from those who do not fit the configuration that is given, which is typically a white, Western, and heteronormative one (see also Chapters 1 and 5). For as Stanfill (2015) reminds us, while users have the option to 'go elsewhere, adapt, or contest this [a site's intended use]', the inbuilt ideal is still always present and 'must be reckoned with' (p. 1961). Crucially, interfaces such as those of cookie consent notices implore people to *do* something and to interact with them. In doing so, they quite literally have to take on the subject position that the interface proposes – ultimately becoming complicit in how they are being stylized by the producers of these texts.

### *6.5.3 Imposed (Inter)Action*

Digital media interfaces impose not only social identities, they also constrain the interactions that may take place. This, I suggest, is linked to how UX writers (and other professionals in software design) design particular interactions through their work. Consider, for instance, the interface of the cookie

**Figure 6.3** Cookie consent notice: Swiss International Air Lines

consent notice in Figure 6.3. What is apparent here is that the interface, though presenting users with a choice of how to act, really only affords one kind of action: to consent to the use of cookies. It is possible to select *which* cookies to allow, but ultimately, as I will discuss below, users cannot choose *not* to allow any cookies. What is more, the multimodal design of this interface very much suggests that one particular action is the preferred response (cf. Sacks, 1987): 'Select all and confirm'. Not only is this option visually the most salient one, it is also the only one that is presented in the typical form of an interface button – and, as such, the one thing that users are most likely to perceive as something that can be clicked in response to this popup (affordances are relational in the sense that people need to perceive them as potential actions, cf. Hutchby, 2001). While the interface does also afford other responses, such as clicking 'confirm selection' to confirm a custom choice, this action is much harder to perceive. In other words, the affordance of these two actions differs.

Importantly, as scholars in HCI have shown, such default and/or pre-selected choices have a significant impact on how people consent to the use of cookies (Singh et al., 2022). In this way, certain designs can manipulate the choice that users make in terms of whether they accept cookies or not (e.g. Utz et al., 2019; Machuletz & Böhme, 2020; Bauer et al., 2021; Bermejo Fernandez et al., 2021). In fact, Machuletz and Böhme (2020) evaluated precisely the impact of 'Select All' buttons, like the one in Figure 6.3, finding that such design choices resulted in a significant increase in acceptance of but not necessarily consent to cookies: while such buttons led to more participants opting-in to all cookies, after the experiment, participants often expressed regret about their choice. From a multimodal perspective, it is arguably the use of salience that contributes to these outcomes. Indeed, I myself instantly clicked the big red button when I first encountered the popup shown in Figure 6.3, not quite realizing what this would actually do, which is to select and confirm *all* four types of cookies and not, as I had assumed, to only confirm the so-called necessary cookies. It is in this way that the interface ultimately imposes a particular action. What this nicely illustrates is how UX writers and their colleagues may work with the (in)visibility (cf. Chapter 5) of particular parts of an interface and how they, in doing so, ultimately impact the actions that users take.

To be clear, UX writers themselves are not oblivious to such problematic aspects of their work. For instance, in a blog article titled 'Cookies UX: just stop the madness', UX writer Rachel McConnell addresses the problematic microcopy of cookie consent notices. Discussing an example similar to the one in Figure 6.3, she criticizes the use of what she calls a 'double bluff':

> Have I accepted or declined? That is the question I ask myself when I see these kind of wordings. The tick boxes aren't pre-selected, and yet I'm asked to *save changes.* I haven't made any changes, is this not the default? If I *accept all cookies* do these become ticked? Gah, my head hurts. (para. 8, emphasis in original)[3]

In many ways, McConnell picks up on the same problematic that I want to unpack. What she identifies is an issue of affordance. Specifically, I suggest that the confusing nature of cookie consent notices that she evaluates with 'Gah, my head hurts' is due to what I would call a *feigned affordance* – something that is suggested by an artefact without actually being possible.

As discussed earlier, the interface in Figure 6.3 makes different kinds of actions possible, which differ in their affordance. Importantly, though, some actions are constrained altogether: it is, for instance, impossible to choose no cookies at all. What is particularly insidious is that the interface suggests that this might be possible. Each of the four types of cookies has a checkbox of the same shape and size. The one for 'Necessary' cookies is selected by default. Users can choose to check and uncheck the other three boxes; however, unlike those, it is impossible to deselect this first box. By using the same visual logic, the interface implies that there is the same action potential, when there is in fact no real affordance at all. It is what I would call a feigned affordance (similar to what Gaver, 1991 calls a 'false affordance'): a deliberately designed and seemingly possible action that turns out to be impossible to take. In this way, affordances act as conditions on say and doability (cf. Maryns & Blommaert, 2002), making some actions possible and others impossible. They constitute the possible uses of an interface as they have been configured by its creators. The way that UX writers craft audiences in the microcopy they write for cookie consent notices hence not only constructs particular participant roles, it can even go so far as to impose particular (inter)actions.

[3] Source: *Cookies UX: just stop the madness.* Retrieved from: https://medium.com/design-warp/cookies-just-stop-the-madness-9da12fcfca94 (accessed 6 October 2022).

### 6.5.4 *Cookie Consent Notices as Symbolic Violence*

Cookie consent notices always position users in particular ways and in particular moments. As I have shown above, more than a matter of audience design (Bell, 1984), UX writers – consciously or not – also invent, stylize, and craft an audience when writing microcopy. Through this, they constrain the social and interactional role that users can take on, imposing both particular actions and particular subject positions onto users. It is in this way that these inconspicuous texts can also exercise a form of symbolic violence. As a way of illustrating this, I offer the cookie consent notice in Figure 6.4. Stylized as a casual interaction, this cookie consent notice, again, does not – or at least not in any direct way – inform users of how this company uses cookies. For this, users would have to read the cookie policy, a text that is often cumbersome to understand and, in part because of this, only rarely read by users (Meier et al., 2020). Instead, the cookie consent notice offers the vague explanation that 'Cookies help us run our services and make them more tasty, so you get a better experience'. At its core, there is a discrepancy between the locution of this text (what the text says) and its illocution (what it is meant to do; cf. Austin, 1975): veiled in humorous wordplay, the text remains unclear about the use and purpose of cookies and simply urges users to click agree. But perhaps most intriguing is the stylized depiction of an actual cookie and the

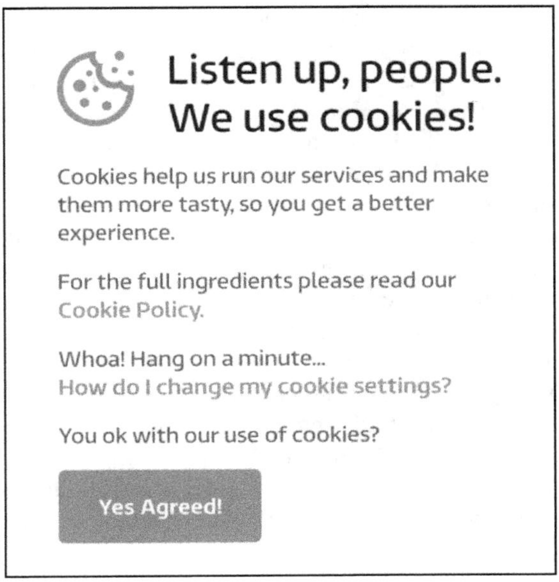


**Figure 6.4** Cookie consent notice: ITV

playful reference to such cookies in the text: cookies supposedly make the company's services 'more tasty', while the cookie policy is said to contain 'the full ingredients'. Such visual and verbal puns are an altogether common practice in cookie consent notices (recall Figure 6.1), but this familiarizing metaphor obfuscates what cookies actually are. Creativity, here, is used to foreground one aspect (humour) while invisibilizing other, perhaps more (business-)critical aspects (the collection of user data). It is quite telling that today, nobody seems to be able to explain or reconstruct *why* cookies are called cookies, though there are at least three different popular origin stories that relate the term to baked goods.[4] Clearly, the metaphor is not self-evident. More importantly, though, it downplays any potential ramifications (recall that there is no mention of what kinds of cookies one agrees to). The use of colloquialisms ('You ok with...?') and emphatic language ('Whoah!'; 'Yes Agreed!') further contributes to this by indexing a casual register. The request to consent is made in such a harmless, friendly way that it is difficult to contest, and yet, ultimately, it constitutes an invisible obligation to click 'Yes Agreed!'. I suggest that this then not only constitutes a conversational inequality (the conversational options available to the company and users are clearly unbalanced) but also a form of symbolic violence (Kramsch, 2021; see also Bourdieu & Wacquant, 1992), whereby users are incited to act in a particular manner that is said to be in their best interest while actually serving the interests of others and violating their privacy.

Following Kramsch (2021), I understand symbolic violence as closely connected to the perlocutionary effect (cf. Austin, 1975) of utterances: it is about getting people to do something, though in such a way that the request is misrecognized as natural and legitimate. Cookie consent notices like the one in Figure 6.4 specifically exploit this; they are framed as harmless, commonplace, and beneficial to the user while exerting subtle pressure (cf. Kramsch, 2021, p. 115) to give consent. This is done through symbolic means; through the linguistic/semiotic form of the cookie consent notice and through the repetitive, or, as Jones (2020a) calls it, iterative, nature of these little texts: 'Every action of clicking "I agree" makes it more likely that I will do the same next time, because "I agree"' has come to be the means by which I can be "on my way"' (p. 94) – in other words, clicking 'agree' becomes part of our habitus. It is precisely in the way that cookie consent notices are framed as a moment of consent that I see their violence: they seemingly give users control but demand acquiescence, resulting in both a loss of agency and a violation of people's privacy in which they are ultimately made complicit. This can thus be understood as a form of symbolic violence whereby companies impose a certain

[4] For an overview of different popular origin stories, see https://cookiecontroller.com/what-are-cookies/ (accessed 25 November 2021)

social order that serves their own interests onto users (cf. Kramsch, 2021). The crux is, as Kramsch (2021) reminds us, that all of this is done under the guise of 'naturalness': apparently, this is simply how digital media work.

As I briefly discussed in the previous section, UX writers are not entirely unaware of the potentially problematic effects of their own work (see McConnell, quoted above, who criticizes the convoluted and manipulative nature of certain cookie consent messages). In fact, during the last stage of my research, I presented parts of my analysis from the current chapter and Chapter 5 at two international UX design conferences. Both times, the audience was very receptive to and interested in my critique. Beyond this, there have also been increasing attempts in the industry itself to engage with such issues of trust and accountability (see, e.g., Wachter-Boettcher, 2017; Monteiro, 2019; again, see also the quote by UX writer Rachel McConnell above). In this way, then, the role of UX writers in perpetuating problematic, hegemonic norms is perhaps not as clear-cut after all. Still, and as I have argued in Chapter 5, UX writers themselves tend to at least partially misrecognize the symbolic power inherent in their work by accepting particular norms and conventions – such as, say, the metaphor of the 'cookie' in cookie consent notices – as natural or normal. It is in these moments that the normative aspects of interface texts are perpetuated unquestioningly. For all their focus on being user centred, then, UX writers themselves, it seems, may not necessarily realize that they are not only writing *for* a particular audience (cf. Massanari, 2010), but that they also craft this audience in and through their work.

## 6.6 Conclusion: The Symbolic Violence of Little Texts

'New media, new audiences?' is the question that Livingstone (1999) asked two decades ago in the inaugural issue of *New Media & Society*. In discussing how audiences might indeed be 'new', she notes the importance of 'the "implied" audience – the audience as presumed, imagined or mythologized' (p. 63), referring primarily to the audiences that are implied in discourses *about* digital media. While my focus here is different, I believe that her words are just as relevant today in our work as scholars of digital discourse. Sociolinguists and discourse scholars have developed useful frameworks for understanding and explaining how different kinds of audiences affect interaction and participation structures, and many have shown that these approaches can also be modified to understand such phenomena in digitally mediated contexts. However, while scholars have examined – and should continue to examine – the kinds of audience roles that people take on in digitally mediated interaction, less attention has been paid to the presumed, imagined, and mythologized audiences that are also present in digital texts: the audiences that are stylized and crafted by professionals like UX writers in the interface design process and how, as a result of this, interfaces construct and constrain the kinds of individuals that may participate in interactions in the first place. (I will elaborate on this in Chapter 7, the concluding chapter of the book.)

In this chapter, I have attempted to sketch a framework for scholars to deal with these other audiences. First, and following Jones (2020b) and Eisenlauer (2014), I have argued that scholars need to account for non-human actors, asking how software interfaces become part of and affect participation structures. Such an approach can help better understand the kinds of interactional conditions that digital media create and how these can lead to interactions that mean one thing to users but another to the technological systems with which users interact. Second, when talking about audiences and digital media, scholars also ought to consider the kinds of ideal addressees that the software interface presupposes. In this regard, I have suggested looking also at the superaddressees (Bakhtin, 1986) that surface in and through interface texts and the processes of user stylization that lead to these superaddressees. Finally, I have explored how software interfaces shape audience and participation through the lens of designed affordances and the imposed interactions that may result from this. This allows an understanding of digitally mediated interaction as determined by user, system, *and* producer (in this case, UX writers). Designed and especially feigned affordances urge scholars to inquire how certain action potentials are built into an interface, thereby shedding light on how interfaces always allow for only certain kinds of legitimate participation and certain kinds of legitimate participants. Ultimately, this shows that participant roles are strongly configured through the interface and the kind of users that producers of digital media presume.

The view of audience and participation that I propose in this chapter is not new, and it is certainly not unique to digital media. Nonetheless, I propose that paying attention to how audience surfaces in this way can be fruitful for a broader understanding of the concept of audience in digital discourse studies. Ultimately, this perspective can show that digital media entail not just audience design but also the more literal design and configuration of audiences. It is in this way that digitally mediated interaction is also centrally shaped by audience *crafting* and the way that professionals such as UX writers encode certain uses and users in the software interface itself – for instance, through the microcopy that they write.

More broadly, my analysis also highlights the significant role of little texts (Pappert & Roth, 2021), which have, at least in Anglophone traditions, been relatively overlooked in discourse studies and sociocultural linguistics.[5] The

[5] A rare exception is Halliday's (1985) appendix on 'the grammar of little texts' in his *Introduction to Functional Grammar*, which is, however, only six pages long and has been removed in more recent editions of the book. By contrast, scholars in German linguistics have, especially recently, started to address such small but important texts in more detail (see, e.g., the edited volume by Pappert & Roth, 2021).

cookie consent notices that I discussed in this chapter are certainly 'little', but, as I hope to have shown, they are by no means inconsequential. As scholars and as ordinary users of digital media, we often misrecognize their importance; however, paying attention to them can reveal much about the symbolic power – and often also symbolic violence – that can be implicated in such little texts. Indeed, it is exactly because they are so small and fleeting that these cookie consent notices can shape digitally mediated interaction in significant ways, impacting the kinds of audiences and participant roles that become possible in the first place. In this vein, my analysis shows that regardless of how people choose to use digital media, the software itself always posits an ideal way of using it; it entails an inbuilt ideal shaped by UX writers and their colleagues that users have to respond to (even if that response is to contest the norm the software produces). Focusing on the production of interface texts thus provides a valuable perspective not just on audience and participation but also on the cultural politics of digital media overall, by theorizing UX writers, software, and users as part of a sociotechnical system rather than as individual, disconnected agents. In Chapter 7, the next and final chapter of this book, I will recapitulate what such an approach to digital media interfaces might entail more generally, making the case for a posthumanist approach to language in digital media that understands interfaces as sites of power where meaning and culture are negotiated.

# Conclusion

# 7 The Cultural Politics of the Interface

## Towards a Posthumanist Sociolinguistics of Digital Media

In 2016, Facebook released the personal data of over 87 million users to Cambridge Analytica, which used the data to support the political campaign of right-wing candidates Ted Cruz and Donald Trump in the 2016 presidential election in the United States.[1] Since the Cambridge Analytica scandal, major software design companies such as Facebook (now Meta), Google, or Amazon have been viewed with a certain amount of mistrust. Their involvement in economic and political scandals has become the norm rather than the exception; it is now quite apparent that these companies are also implicated in social, economic, and political systems of control. Notwithstanding such overt examples of the politics and power of software companies, what I hope to have shown throughout this book is that these companies also exert power in more subtle ways: through the technological artefacts they produce and the cultural values that are articulated through these artefacts. My main concern has been to document how these sociocultural politics surface in the production of digital media and specifically in the work of UX writers. I have shown, for example, that the interfaces which UX writers produce are inherently and unavoidably normative in at least two ways: on the one hand, as texts that can reiterate existing, ideological views but, on the other hand, also as embodied experiences, whereby the interaction between technology and user reproduces hegemonic norms. It is this second normativity that I want to explore further in my conclusion here, suggesting that a posthumanist approach to digital media interfaces can help scholars produce a more complete account of the way power operates in and through contemporary digital media.

As a way to establish what I mean by a posthumanist approach to digital media interfaces, I return to Facebook (now Meta), specifically, to the visual and linguistic design of Facebook's status update interface, as shown in Figure 7.1. Along with written status updates, the interface allows users to select an emoticon to indicate their current mood. While most of these

[1] Source: *Revealed: 50 million Facebook profiles harvested for Cambridge Analytica in major data breach.* Retrieved from www.theguardian.com/news/2018/mar/17/cambridge-analytica-facebook-influence-us-election (accessed 11 October 2022).

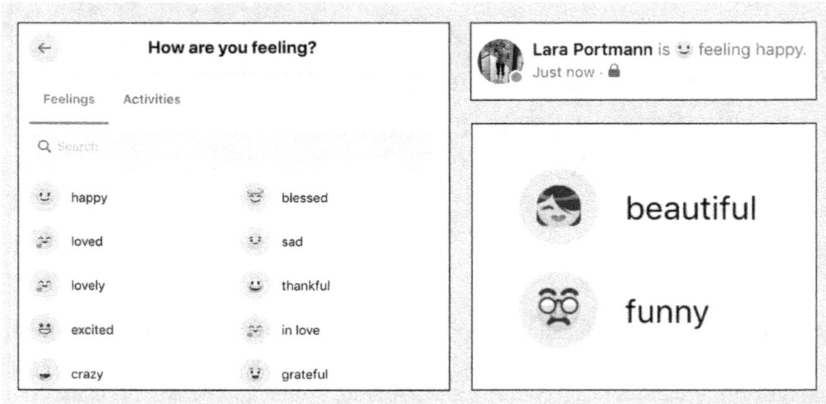


**Figure 7.1** 'Feelings' in Facebook's status update interface

emoticons are gender-neutral (some are even objects), two of them stand out in particular due to their gendered design: those labelled 'beautiful' and 'funny'. 'Beautiful' is indexed through a decidedly feminized representation with a bob haircut, red lips, and long, black eyelashes. 'Funny', on the other hand, is indexed through bushy eyebrows, glasses, a pronounced nose, and a moustache. Presumably, this is a reference to the American comedian Groucho Marx, and, therefore, an ethnocentric design choice. At the same time, 'funny' is of course also subtly encoded as male. The emoticons remain the same regardless of the gender or preferred pronoun users choose in their profile settings. Clearly, these emoticons are neither random nor neutral, instead perpetuating that 'funny' is something men and 'beautiful' something women are (or can be). Facebook's status update interface therefore reinforces gendered stereotypes. More importantly, though, it makes anyone who uses these emoticons complicit in perpetuating these stereotypes. If I want to express in my status update that I feel 'beautiful', I cannot do so without making reference to a very specific, heteronormative idea of beauty, where beauty is expressed through a particularly normative notion of femininity. Small interface moments like this significantly shape and sometimes literally determine what can be said and how things can be said by people using digital media.

That interfaces are normative is certainly not a new claim (e.g. Selfe & Selfe, 1994; Djonov & van Leeuwen, 2017). Having said which, current approaches in digital discourse studies tend to address the normativity of interfaces primarily by examining interfaces as textual artefacts. Based on my own research, however, I would like to make the case for extending this by attending to the more-than-representational aspects of interfaces and to how interfaces exert their power not just through textual and discursive aspects but also – or especially – through the way they incite users to act upon and through them. In what follows, I first offer a brief summary of the five analytical chapters of this book and consider how my findings relate to matters of status, power, and normativity. My main point is simple: in order to understand the cultural politics of digital media, it is necessary to examine not just the use of digital media or digital media as such but to also investigate the production processes that shape them. Such a perspective allows for a more complete picture of the social and semiotic norms that come to be encoded in digital media (cf. Djonov & van Leeuwen, 2017) and that influence what users (can) do with these technologies.

At the same time, turning to the production of digital media and its producers opens up possibilities for a potential dialogue between researchers and practitioners. It is with this in mind that in the second part of the conclusion, I return to UX writers' own understandings of language and communication that I have teased out in the preceding chapters in order to ask what it is that sociocultural linguists might perhaps learn – or at least appreciate – from UX writers. Concretely, I suggest that UX writers are particularly attuned to the ways digital media interfaces operate not just through what is said but, above all, through the action(s) they incite. In light of this, I propose that the expertise of UX writers can offer a starting point for extending the recent posthumanist turn in sociocultural linguistics (Pennycook, 2018a) with regards to digital media. In the final section, I reflect on what such a posthumanist sociolinguistics of digital media might look like and how it can be used to examine how the work of companies like Google or Meta is political not just when they are involved in literal politics but also when people use such software for everyday activities – like posting a status update on their Facebook page.

## 7.1 Summary: Status, Power, and Normativity in UX Writing

While digital discourse scholars have conducted substantial research on how people communicate with and through digital media (e.g. Thurlow & Mroczek, 2011; Jones et al., 2015; Georgakopoulou & Spilioti, 2016), little attention has been paid to the role of the producers of these media and how people like UX writers shape the way people (can) communicate. Some might argue that what producers intend does not matter and that what counts is what users end up doing with digital media. Yet, as I have argued in this book, regardless of how people choose to use digital media, the software itself always posits an ideal way of using it (cf. Chapter 6). Stanfill (2015) thus suggests that '[t]he interface makes a normative claim' (p. 1061) – it entails an inbuilt ideal users have to respond to, even if that response is to contest the norm the software produces. Importantly, and as I hope to have shown through this book, software interfaces and their normative claims are not coincidental; rather, they are purposefully designed by people like UX writers. In this way, I have sought to unpack both the practical and ideological implications of UX writing, showing that focusing on the production of interface texts can provide a valuable perspective on the conditions of UX writing as elite language work but ultimately also on the broader cultural politics of digital media.

To examine the connection between UX writers' elite language work and the cultural politics of software interfaces, I began in Chapter 2 by mapping the (language) work and expertise of UX writers. In this regard, I highlighted some of the key ways in which UX writing is connected to other professions in software design, most notably to user experience design and human-computer interaction (HCI). This, in turn, impacts how UX writers approach language in their own work: they tend to view microcopy (i.e. interface texts) not necessarily as something users are meant to read but more so as something users ought to act upon. What started to emerge is that UX writers are very much concerned with what is going on besides language and with the overall outcomes or effects of language. More broadly, I illustrated in Chapter 2 that UX writers can be understood to work as wordsmiths (Thurlow, 2020c); they are markedly high-end, relatively elite language workers whose work is valued highly, both in symbolic and in material terms.

In the second part of my book (Chapters 3 and 4), I explored UX writers' status as wordsmiths in more depth, asking in particular how they negotiate their position in the local and wider hierarchies of (language) work. In Chapter 3, I looked at this by focusing on skill discourse, discussing the construction and codification of writing-as-designing as a unique skill that UX writers claim in order to establish the (more) privileged status of their language work. As such, UX writers typically frame their work as being concerned with the communicative function of words rather than their verbal form. I explored this (ideological) distinction between form and function further in Chapter 4, where I discussed how UX writers (dis)avow creativity. Again, my primary focus in Chapter 4 was on UX writers' status claims and on how they establish and maintain their position as relatively elite language workers. In this regard, I suggested that claiming non-creativity is another way for UX writers to establish the boundaries of their professional field, all while framing their own work in a positive light. As a whole, Chapters 3 and 4 highlighted UX writers' strong orientation to design as a professional field, and with this, the professional and also cultural norms of UX design – norms that, as I showed in Chapter 5, can be both deep-seated and problematic. These two chapters thus showed that status and normativity are closely connected in UX writing. It is UX writers' elite position that grants them status; at the same time, their elite position at least partially depends on accepting and reproducing existing, hegemonic norms that persist in UX and software design.

In the third part of my book (Chapters 5 and 6), I turned to discussing the normative consequences of UX writers' work in more detail. In Chapter 5, I examined the ideology of the invisible interface in UX writing. Central to their status as wordsmiths, UX writers' (supposedly) invisible language work has, I argued, wider, cultural-political consequences. The invisibility UX writers strive to produce depends on privileging a particular, normative position from which things are rendered (in)visible. In this way, UX writers approach and imbue digital media with ideal or idealized uses and users. This was the focus of Chapter 6, where I discussed the normative impact of UX writers' work in terms of how UX writers ultimately engage in inventing, stylizing, and crafting particular audiences. As such, this final analytic chapter underscored the normativity of interfaces; they are normative not just as texts that reproduce ideological views but, above all, through the dynamic (inter)actions they incite.

Across all the chapters, I have been particularly interested in UX writers' own understanding of language and communication and the beliefs they hold about how language works – or is supposed to work – in digital media interfaces. Ultimately, I have shown that examining UX writers' views of language is central to understanding both their role as wordsmiths and the cultural-political implications of their work. On this basis, I now want to reconsider how UX writers approach language; although this time, I am not concerned with the normative implications of UX writers' semiotic ideologies, but rather with the possibility that sociocultural linguists might learn something from UX writers.

## 7.2 The Role of Wordsmiths and Language Scholars

In this book, I have written about UX writing from an academic perspective; as a former practitioner, however, my interest in UX writing has always been twofold. On the one hand, and as an academic, I have been keen to document this work from a critical perspective and to understand its particular, cultural-normative impact. On the other hand, and as someone who is still actively involved in the UX writing community, I am also interested in what my research might mean to UX writers themselves and how they might respond to it. In this regard, I want to briefly consider how these two perspectives could be integrated. This brings me full circle to the question of how sociocultural linguists might look at UX writers not just as language workers but also as language experts in their own right.

In reflecting on the differences between how practitioners and academics might think differently (or not) about language, I find myself drawn to the work of scholars like Macgilchrist (2017, 2020), who demonstrates how discourse ethnographic approaches can create not only novel theoretical insights but may also constitute a form of knowledge exchange between academics and other language experts. The different conditions under which wordsmiths and sociocultural linguists conduct their work are at the heart of Macgilchrist's (2020) work. Discussing the work of history textbook authors, she reflects on how the educational publishing industry as an institution that governs what counts as worth knowing affects the language work of these authors. In her case study on (post)colonial discourse in textbooks, Macgilchrist highlights that the particular knowledge textbook authors have of the publishing apparatus and of schools themselves ultimately makes them better at formulating a postcolonially oriented textbook than discourse scholars. She shows that the authors' work, which from a text-analytic perspective may not seem particularly progressive, is in fact more likely to effect postcolonially oriented change, due to it being more 'teachable' than a progressive text. Here, it is the concrete expertise of applied language experts that turns out to be at least as valuable as that of sociocultural linguists.

Especially inspiring is Macgilchrist's (2020) ethical commitment to remaining critical of the work of textbook authors while also acknowledging these wordsmiths as experts themselves. In this regard, Macgilchrist makes a case for replacing the all-too-comfortable academic *but* with a Deleuzian *and*, holding space for academic critique *and also* for the expert knowledge of wordsmiths themselves. Instead of hedging the positive insights of wordsmiths through the use of *but* (e.g. they may have applied knowledge, but they still reproduce normative discourses), Macgilchrist integrates the expertise of wordsmiths with the critical thinking of academia, seeking to hold both perspectives together and recognizing them as equal. Similarly, several case studies in Perrin and Kleinberger (2017) point to the value that applied linguistic research has when engaging other language professions and how 'research in the workplace' (Ehrensberger-Dow et al., 2017) can lead to mutual learning between academics and professionals; while scholars in applied linguistics (e.g. Candlin & Sarangi, 2003; Sarangi, 2005) have long argued against treating (language) workers only as objects of study rather than as sources of valid linguistic insight.

Meanwhile, Thurlow and Britain's (2020) recent work on dialect coaches is exemplary in showing how the work of wordsmiths can help academics revisit their own conceptualizations of language and communication. In particular, Thurlow and Britain are keen to not simply critique wordsmiths' work but to be open to learning from them, acknowledging that academic perspectives on dialects are neither better nor more objective than those of dialect coaches (cf. also Irvine & Gal, 2000). As Thurlow and Britain note, these differences are in part due to the somewhat different objectives of wordsmiths and academics: while wordsmiths are usually concerned with specific, individual cases, language scholars are typically expected to offer a (more) generalized argument.

How language is conceptualized thus depends on the particular conditions under which it is imagined (cf. Urciuoli, 2016). In other words, the material and structural conditions of practitioners' and academics' work lead to a somewhat different but perhaps not incompatible understanding of language. It is with this in mind that I want to offer some final remarks on the language work of UX writers, reflecting on the connections (rather than differences) between 'their' work and 'ours'.

The way UX writers think about language is not necessarily so different from that of sociocultural linguists. For instance, how they understand matters such as multimodality (cf. Chapters 2 and 5) in fact aligns quite well with current scholarly approaches. (I will comment on this in more depth in Section 7.3.) In thinking about the cultural-political consequences of UX writers' language work, I am equally reminded that while UX writers may be less aware of the ideological and normative implications of their work, that is not to say that they are unaware of them altogether. For instance, one of my interviewees remarked that the meaning of the word 'share' has changed in the wake of digital/social media, with 'sharing' now referring to the sharing or publishing of information more so than communal sharing. This is, of course, the central argument of media and communication scholar John (2017) in his recent book *The Age of Sharing*. Another interviewee pointed out the significant influence Silicon Valley culture has on whether (and which) words and discourses are circulated globally; others yet considered how UX writing shapes people's experiences in terms of social in- and exclusion (for instance, through gendered language), or they reflected on the role UX writers can play in abolishing racist terminology (such as the still prevalent use of 'master/slave' or 'blacklist/whitelist' in software engineering practices). More generally, there has been a shift in UX design as a whole in recent years. Topics such as sociocultural impact, ethics, and UX practitioners' responsibility in shaping cultural discourses have increasingly become prime topics at industry conferences; these matters are also publicly discussed in podcasts and blog posts produced and circulated by industry practitioners, including UX writers.[2]

There is necessarily a difference in how much space matters of language and cultural discourses take up in the work of UX writers and in that of sociocultural linguists. Engaging with such matters once or twice a year at a conference is certainly not the same as doing so on an ongoing basis in one's day-to-day

[2] Needless to say, this observation is based on my own experience in the professional field. At the professional conferences I attended between 2017 and 2022, there were usually several presentations that engaged with the sociocultural impact of UX design (though, of course, not necessarily under that label). Similarly, conference calls have increasingly emphasized these matters with conference topics such as 'Impact' (*EuroIA 2019*), 'Trust, ethics and integrity' (*World Usability Day 2022*), and 'Establishing design ethics' (*Interaction 23*).

work. Evidently, this is tied again to the different structural conditions of practitioners and academics. Language scholars, especially those working in fields like critical sociolinguistics or critical discourse studies, are paid to think about such matters. UX writers, of course, are not; they are primarily paid to produce writing that 'works'. Perhaps, then, the difference in 'our' and 'their' thinking is not so much a qualitative but a quantitative one. It is certainly true that the conditions of UX writers' work do not allow them to spend as much time thinking about cultural-political aspects as sociocultural linguists can, but that does not mean that they never think about these matters. Given this, it makes all the more sense to move beyond the 'academic but' (Macgilchrist, 2020), which necessarily hedges and de-valorizes the insights of non-academic language experts like UX writers.

Indeed, perhaps it is most productive to think of these wordsmiths not just as research informants but also as valid sources of sociolinguistic insight or even theory (cf. Sarangi, 2005). This is not to say that UX writers themselves engage in such theorizing (not in the same way sociocultural linguists would); what I am suggesting is rather that respecting the work and expertise of these practitioners can offer new opportunities for expanding sociolinguistic theory. In the second part of this conclusion, I now turn to exemplifying this mindset by considering how the expertise of UX writers can help expand the recent turn to posthumanism in sociocultural linguistics.

## 7.3 Learning from UX Writers: Transmodality and Embodiment

In turning to a posthumanist perspective, I start by considering two concepts UX writers employ in their work, but which have received, until only recently, slightly less attention from sociocultural linguists (especially from those working in digital discourse studies): transmodality and embodiment. These two concepts have surfaced in the previous chapters as core approaches of how UX writers theorize language and communication. With regards to the first, transmodality, I suggest that UX writers are especially good at understanding how semiotic modes are interrelated. While multimodality is certainly not a new topic in sociocultural linguistics, I propose that the way UX writers approach this in their work entails a deeper engagement with, and understanding of, transmodality and transduction (Bezemer & Kress, 2008; see also Thurlow & Jaworski, 2017b). Furthermore, and turning to embodiment, UX writers are also particularly well-attuned to non- or more-than-representational aspects of communication (cf. Lorimer, 2005; Thrift, 2007) and the way that texts are not just read but experienced by readers in an embodied way (cf. Jones, 2022a). These two issues, transmodality and embodiment, are connected insofar as they both decentre linguistic and textual aspects; it is in this way that they ultimately serve as a starting point for developing a posthumanist approach to digital media.

### *7.3.1 Transmodality*

As I have shown throughout my book, UX writers understand communication as inherently multimodal (see especially Chapters 2 and 5). To them, separating the linguistic or verbal aspects of an interface from its visual design seems counterproductive, if not impossible. While multimodality studies have led sociocultural linguists to think beyond the written and spoken word, the field's primary focus is, for better or worse, still language. Consequently – and in contrast to how UX writers approach language – sociocultural linguists are prone to separating out the linguistic aspects of communication at the expense of what is going on besides language (see Thurlow, 2016). Given that their primary focus tends to be language, it is perhaps not surprising that sociocultural linguists might struggle with examining what lies besides language. This is, however, exactly what UX writers excel in. Language is also their key concern; yet, they understand well that communication is always more than words and that sometimes, words are not the point at all. As I discussed in Chapter 5, they certainly understand that language cannot be separated from other semiotic modes. To illustrate and extend this idea, I offer Extracts 7.1 and 7.2, where two different UX writers describe how they think about and incorporate other modes in their work processes. Both Extracts are taken from my own interviews with UX writers. In the first instance (Extract 7.1), Mona, who has been working as a UX writer for about ten years, talks about how the visual design of an interface may influence her writing.

**Extract 7.1: Mona on transmodality in UX writing**

1. Here are three possible palettes on how we can approach it [the design]. . . . So, I
2. came in and was like, "Okay let me let me design language around each of those."
3. So, you know, this one has sharper corners, let's use a more, you know, staccato
4. kind of language. This one has a softer feel to it, so let's use more descriptive
5. language. . . . It's just like anything else, it has a tone, it has a feel.

In Extract 7.1, Mona refers to her practice as 'design[ing] language around' the visual aesthetic of the interface (line 2). As such, the visual design of the software appears to centrally shape the style of her writing. In her explanation of how the visual and verbal aspects of the interface are connected, Mona invokes even more modes: note how she uses musical terminology ('let's use a more . . . staccato kind of language', lines 3–4) and refers to touch ('this one has a softer feel to it', line 4) to describe her work. To scholars of language, Mona's description may seem somewhat trivial. Certainly, sociocultural linguists have developed different and perhaps more nuanced metalanguage for describing these phenomena. A 'staccato kind of language' (lines 3–4) might be described as parataxis, characterized by short sentences without subordinate clauses, while language that 'has a softer feel' (line 4) might entail more voiced sounds or rounded graphemes, which tend to be associated with round shapes (cf. Ramachandran & Hubbard, 2001; see also Cuskley et al., 2017 on orthographic and phonological aspects of sound symbolism). Regardless of this, though, Mona's approach to writing clearly goes beyond considering only linguistic aspects. To her, UX writing appears to be an inherently and necessarily transmodal matter.

As a second example of UX writers' transmodal approach to their work, I offer Extract 7.2 from my interview with Celia, an experienced UX writer working at a major international software company in the United States. Here, she responds to my question about what UX writing generally entails. Like Mona, she seems to understand her work as inherently transmodal, and she appears to be particularly attuned to the interplay of various modes.

**Extract 7.2: Celia on transmodality in UX writing**

Say, it's a login screen and there's, like, one field and then there's a second field below it and a big button. I don't think many people read what's the label on that field and what's the label on that button. They're just like, "Oh it's 'username' and 'password'", and then they do it. . . . We have to be aware of what are the times when the visuals are going to make it impossible for our copy to do anything. . . . And that's not always a weakness, right, sometimes it's a strength, sometimes the job is about "Hey, I've see– I think I know how this screen should look just because I know what people will expect to see here." . . . It is not just did we craft strings[3] that get the message across, but are those strings being presented in a way that they get a chance to do that.

In this instance, Celia explains how regardless of the words she writes, sometimes the people using the interface will not read these words because they are too focused on other modes: sometimes 'the visuals are going to make it impossible for our copy to do anything' (line 5). As a consequence, Celia's approach to writing microcopy has to be a transmodal one from the start. She has to consider not just what to write but also how the writing will be presented alongside other modes: '[i]t is not just did we craft strings that get the message across, but are those strings being presented in a way that they get a chance to do that' (lines 8–10). More than this, Celia appears to not just consider the influence of the visual design on her writing but to actively make use of the interplay between various modes: she claims that the dominance of the visual mode may also be a strength she can exploit (line 6). Sometimes, then, the work of a UX writer is not about writing specific words, but rather about understanding when the words don't matter. UX writing is therefore based on

[3] In simplified terms, a string is a data type used in programming to represent text. UX writers generally use the term 'string' to refer to an individual section of text: for instance, the text on a button may be one string, while a sentence written below it would be stored as a second, separate string.

the fundamental understanding that language functions not in isolation but as part of an assemblage of modes. This is, of course, a truism in sociocultural linguistics nowadays, but my research shows how practitioners (already) work with the same fundamental understanding.

What Extracts 7.1 and 7.2 reveal is that UX writers seem to be good at understanding both the 'division of labour' (Kress, 2010) between different semiotic modes as well as the relationships between modes, how they may supplement one another (as in Mona's case) or work differently or even in opposition (as in Celia's case). Especially intriguing, however, is how these UX writers approach transmodality not simply as a combination of individual, separate modes (as sociocultural linguists tend to do); rather, they understand texts – in this case, interfaces – as an assemblage of modes that are inherently intertwined and interdependent. In simple terms, they understand that the sum is more than its parts. This difference between 'our' and 'their' approach is not surprising; while sociocultural linguists may be interested in disentangling texts in order to understand how they work, UX writers are primarily concerned with putting these texts together – a task where separating out language from other modes is likely counterproductive.

Despite the different goals that UX writers and language scholars may have, though, sometimes, it is exactly the *synaesthetic* (Kress, 1997) interplay between different semiotic modes that matters most, also for sociocultural linguists. In this regard, Thurlow and Jaworski (2017b) illustrate that approaching texts as semiotic assemblages can extend scholars' understanding of the cultural work and impact of such texts (see also Jaworski, 2017; Thurlow, 2015, 2020d). As a case in point, they show, for instance, that things such as a plastic frequent-flyer membership card acquire their meaning only when one considers them as a transmodal practice – it is through the interplay of words, colours, and materials that the card *becomes* a membership card instead of a just a worthless piece of plastic. Similarly, UX writers understand that language can produce a particular effect, for instance, a 'softer feel' (Extract 7.1, line 4), not alongside but only in combination with other modes.

UX writers also understand well how different modes may influence or condition one another. As Celia (Extract 7.2) notes, depending on the look of an interface, users may be more or less likely to read the individual words in that interface. Users may recognize a login interface, Celia explains, not because they see the words *login*, *username,* or *password*, but rather because they know where (or when) they can expect that interface, and because they perceive its overall layout: 'there's . . . one field and then there's a second field below and a big button' (lines 1–2). What Celia recognizes is that these layout practices condition the interface as a transmodal text: one mode (layout) determines how another mode (written language) can or will be perceived.

Working with another applied context, Murphy (2012) discusses the way different semiotic modes relate to and condition one another as an aspect of transmodality. Analyzing the day-to-day interactions of two furniture designers as they work on completing a new assignment (the design of a candleholder), Murphy observes that the design and physical form of the candleholder is shaped by sketches and computerized 3D drawings but also by words (e.g. 'soft and nice') and bodily movements (such as a sloping gesture). Importantly, he shows how actions carried out in one mode shape subsequent actions in other modes. A round and sloping gesture may become a sloping shape in the computer drawing, with the gesture thus conditioning the shape of the 3D model. This leads Murphy to argue that modes are not simply layered on top of each other; rather, they are interdependent. Different modes influence one another as well as any subsequent communicative action;, they 'perforate and interpenetrate each other' (p. 1969) through what Murphy (2012) calls *semiotic modulations*. Relatedly, Goodwin and Alim (2010) speak of *transmodal stylization* in order to theorize how sociolinguistic style may be produced in a way that 'cuts across various modalities' (p. 191), by which they mean that multiple, possibly opposed, social meanings may be expressed simultaneously through the deployment of more than one mode (e.g. gesture and spoken language) that each index a different sociolinguistic style.

Writing from an explicitly social semiotic perspective, Jaworski (2017) similarly speaks of the 'moiré-like' nature of the interplay between semiotic modes. 'Moiré', in this case, refers to the layering of particular semiotic patterns on top of each other, whereby different modes 'fluctuate in a constant ebb and flow of interactional synchronicity and rupture' (Jaworski, 2017, p. 535). As such, different modes are understood as only ever partially responsible for the interaction that takes place. All of this points to a more dynamic view of meaning making.

Approaches emphasizing transmodality are of course not entirely new in sociocultural linguistics and have, for instance, been discussed under the rubric of *synaesthesia* (Kress, 1997) or *resemiotization* (Iedema, 2003). In this regard, Kress (1997) asserts that multimodality (or what more recently has been called transmodality) always entails the complementary processes of transformation and transduction, that is, the reshaping of semiotic resources within and across modes, respectively. Such reshaping of meaning leads to what Kress calls *synaethesia*, whereby something that has been translated from one mode into another leads to qualitatively new forms of meaning. Relatedly, Iedema (2003) proposes the concept of *resemiotization* in order to theorize how semiotic resources are translated from one mode into another as social processes unfold, arguing that meaning making is an inevitably dynamic and socially situated process. More recently, scholars have discussed the relationships between modes also under the label *intermodality* (cf. Stöckl, 2014, on intermodal harmony; see also papers in Maier, 2017). In general, however, the dynamic interplay between modes is still only occasionally addressed in sociocultural linguistics. Contrary to this, such *transmodality* is firmly rooted in the work of UX writers, who understand that modes other than written language may well determine not only how but also if people will read their writing.

### *7.3.2 Embodiment*

UX writers are particularly attuned not just to the transmodal nature of interface texts but also to their more-than-representational aspects (Lorimer, 2005). As Extract 7.2 starts to show, UX writers are well aware of the dynamic, situated nature of software interfaces, recognizing that users tend to fluidly click through interfaces and thus understand interfaces as one single interaction rather than as individual, separate texts. It is this which brings me to a second aspect that is particularly prominent in UX writers' work: embodiment. Unlike many language and discourse scholars, who have, as Thurlow (2016) points out, often centred textual aspects of communication and paid only 'little attention to the embodied, intuitive, affective ways of doing and knowing' (p. 507), UX writers themselves frequently place the embodied experience of texts at the centre of their work. In this regard, I offer Extract 7.3 from my interview with Miriam. At the time of our interview, Miriam had been working as a UX writer for seven years; here, she explains how she usually processes a new assignment.

**Extract 7.3: Miriam on embodied experience in UX writing**

1. We try to figure out what we want to say, how we want to, how we want to say it
2. . . . you know, (we) try to get the right experience. And my tools are the words, but,
3. like, everything combined with visuals, with interaction creates this experience.

What is particularly apparent in this case is how Miriam's understanding of communication is decidedly more-than-representational. Naturally, Miriam's work as a UX writer entails making decisions about the words that will be used in an interface and 'figure[ing] out what we want to say, how we want to . . . say it' (line 1). She suggests, though, that what ultimately matters is 'everything combined with visuals, with interaction [which] creates this experience' (line 3). Like Mona and Celia (Extracts 7.1 and 7.2), Miriam thus understands her work as inherently transmodal; and what she sees as the outcome or goal of her work is not an interface, but an 'experience' (lines 2 and 3).

I have commented on UX writers' use of the word experience in Chapter 5, identifying it as a key organizing principle that UX writers use to describe their work. The term *experience*, I argued, indexes UX writers' orientation towards the field of UX design, but it also erases or obscures the materiality of digital media. Here, I am interested primarily in how this focus on something more-than-representational also influences UX writers' understanding of language, and, specifically, how it is connected to an embodied understanding of communication. Notice, for instance, how Miriam singles out 'interaction' (line 3) as a key aspect of the experience she and her colleagues are seeking to create – evidently, she sees this as an important aspect of how interfaces work. In doing so, she again draws attention to the dynamic nature of the interface (cf. Chapter 2); this, I propose, is connected also to an understanding of communication as inherently embodied (cf. Bucholtz & Hall, 2016), as something people *do* with all their senses.

To illustrate UX writers' more-than-representational and embodied approach to interfaces, I offer another example (Extract 7.4), this time from my interview with Tina, a senior UX writer working at a large, internationally known software company. In this instance, Tina talks about her general approach to (UX) writing.

**Extract 7.4: Tina on embodied experience in UX writing**

1. Besides language, there is the positioning of the language, where that language
2. appears, if it appears at all. . . . I had a designer reach out a couple days ago who
3. said, "I need, I need a piece of copy here", and I looked at the page and I realized
4. there's no copy needed. Like it's already– it's clear to the user where they are, all
5. they need is the button to get to the next page.

Once again, Tina appears to be very aware of how (written) language intersects with other modes and that owing to this, sometimes, 'there's no copy needed' (line 4). She also addresses that writing always entails more than just the selection of particular words: 'where that language appears [and] if that language appears at all' (lines 1–2) matters just as much. In this way, Tina recognizes that meaning is always also linked to the compositional and the interpersonal (meta)function of language (cf. Halliday, 1978; Kress & van Leeuwen, 2001). What interests me most about Tina's account, however, is how she discusses the interface as a dynamic object that is made up of more than individual screens. Tina's decisions about what to write (and whether she should write anything at all) appear to be made primarily based on users' (imagined) movement through the interface. As such, Tina takes a decidedly embodied perspective on interface texts: she spatially locates users in the interface ('it's clear to the user where they are', line 5). In fact, she goes on to explicitly characterize users' engagement with the interface as a form of movement by suggesting that users need to 'get to the next page' (lines 5–6). Tina's approach to (UX) writing thus highlights the fact that users do not read interface texts in isolation; rather, Tina illustrates how users engage in a more *embodied* form of using the interface that entails the physical clicking or touching of buttons (line 5) and a dynamic movement through the pages of the interface.

In sociocultural linguistics, embodiment is still a relatively recent topic of interest (cf. Bucholtz & Hall, 2016; see also Thurlow, 2016), despite the fact that it is, as Bucholtz and Hall (2016) point out, central to both the production and reception of language: speaking, listening, reading, and writing all depend on people's use of their bodies. Notwithstanding, the human body is increasingly foregrounded in certain areas of sociocultural linguistics. For example, scholars in mediated discourse analysis have long centred the role of the historical body and the way discourse is constituted also through the embodied habitus of people (e.g. Scollon & Scollon, 2004; Norris & Jones, 2005);[4] conversation analysts have centrally investigated the use of the body in interaction (e.g. C. Goodwin, 2007; Mondada, 2011; Cekaite & Mondada, 2020); and there is a growing body of research examining matters of language and gender from an explicitly embodied perspective (e.g. Peck & Stroud, 2015; Hall et al., 2020). However, discourse-focused scholarship has, as Thurlow (2016) observes, tended to fetishize the text and centre the textual and representational. This is perhaps particularly true for digital discourse studies, despite the fact that, as Jones (2022b) recently pointed out, the body is increasingly reasserting itself in digital communication, especially through new forms of 'synthetic embodiment' such as digitally enhanced Snapchat pictures or Apple's Animojis, which transpose the recording of a users' face onto an animated emoji. In this sense, I see embodiment not as a wholly new concern but as a somewhat neglected perspective in sociolinguistic research on digital media.

Writing about digital literacy, Jones (2022a) suggests that an often overlooked dimension when it comes to digital media is their material qualities and the fact that digital media are always experienced as embodied practices. This is not to suggest that there is no research attending to the embodied nature of how users engage with digital media; in fact, Jones's own research (e.g. Jones, 2011, 2015, 2020c) is exemplary in addressing the entanglement of human bodies and technologies in digitally mediated interaction. In this regard, Jones discusses, for instance, the entextualization of the human body in amateur skateboarding videos (Jones, 2011) or through self-tracking apps (Jones, 2015); more recently, he has also addressed the role of embodiment with regards to visual digital practices (Jones, 2020c). Similarly, Keating's (e.g. 2005; Keating & Mirus, 2003) earlier work on sign language users in video-mediated online interactions highlights the intersections of bodies and technologies, just as research on language and sex/uality online has likewise tended to centre matters of embodiment (e.g. Adams-Thies, 2012; Varis,

[4] To be clear, the *historical body* in mediated discourse analysis refers not necessarily to people's physical body but to the specific knowledge and practices that are embodied within particular social actors, similar to Bourdieu's (1990) concept of habitus.

2021; see also Jones, 2009). Crucially, research like this shows how it is never just the body that is being represented digitally; rather, the material (the body) and the digital (the interface) mutually affect and constitute one another.

Despite some digital discourse scholars' engagements with embodiment, when it comes to interfaces in particular, scholars still tend to often neglect the fact that interfaces are something people also physically and tangibly click through; instead, scholars prefer to discuss interfaces as static, bounded entities. However, approaching interfaces as something embodied might allow scholars to better understand how they function not just on the level of representation but also in terms of the patterns of usage they create, that is to say, how their 'eventfulness' (Dieter, 2015) can encourage specific, cultural-normative performances. A case in point is Friz and Gehl's (2016) study of Pinterest's sign-up interface: discussing the sign-up process as a 'gender script', the authors show how it is not an individual screen that makes Pinterest feminine-coded; rather, it is through the way that Pinterest guides new users through their first steps on the platform that feminine-coded gender performances are encouraged (such as the act of curating – rather than creating – content). The interface, through its suggested and oftentimes required processes and actions, thus shapes both the embodied and discursive actions of users. Ultimately, approaches that attend to the dynamic, situated, and embodied aspects of digital media interfaces encourage researchers to address interfaces not just as texts but also as lived experience (cf. Jones, 2022a). In the end, users literally move parts of their body (if only their fingers) to manipulate the interface; interface design choices also pre-scribe the embodied actions of users. For sociocultural linguists, this likely means developing a less representational approach to language in digital media interfaces that conceptualizes how users engage with interfaces as an embodied practice – similar to the way UX writers conceive of it.

Transmodality and embodiment are connected insofar as they both decentre the linguistic and the textual. It is not surprising that UX writers are particularly good at this: owing both to the distributed nature of their work (cf. Chapter 2) and their relative position as (language) workers vis-à-vis their colleagues (cf. Chapters 3 and 5), UX writers always need to understand not just language but also what is going on beside language. This includes also an understanding of interface practices as embodied, and of texts and bodies as inherently connected. Ultimately, then, as I have teased out in this book, UX writers work in a posthumanist way. Their practice might be seen as sitting at the cutting edge of sociolinguistic theory; it is in this way that taking UX writers' work and expertise seriously offers an opportunity for expanding current posthumanist approaches in sociocultural linguistics. Specifically, I propose that the work of UX writers shows that the study of interfaces demands a posthumanist lens, indicating that scholars ought to engage more with the more-than-representational, unstable, and dynamic nature of the software interface in order to address its normative potential. In the last part of this conclusion, I offer a preliminary outline of what such a posthumanist approach to digital media might look like.

## 7.4 Towards a Posthumanist Sociolinguistics of Digital Media

Posthumanism is a broad term that is used to describe a range of allied but also divergent contemporary movements in the humanities and the humanistic social sciences. My own use of the term, as I have outlined in previous areas of this book (see, especially, Chapter 1), orients primarily to questioning the notion of the (rational) 'human' as the sole arbiter of legitimate action (e.g. Herbrechter et al., 2020). Posthumanist thinking is certainly not new; however, it is only very recently that it has been taken up in language and discourse studies; in this case, scholars have begun to consider how non-human and non-linguistic aspects shape discursive practices (e.g. Pennycook, 2018a; Lamb & Higgins, 2020; O'Halloran, 2022). This posthumanist turn in language scholarship has been explored most prominently by Pennycook (2018a), who posits that 'there is no longer a world "out there" separate from humans and represented in language but rather a dynamic interrelationship between different materialities' (p. 449). In this regard, Pennycook understands meaning making as taking place not just through language and discourse but also through people's being-in-the-world. This, of course, aligns nicely with non-representational theory (cf. Thrift, 2007). Pennycook discusses a variety of discursive phenomena from this perspective, perhaps most notable is his revision of the linguistic notion of *repertoire*, which he conceptualizes as 'an emergent property deriving from the interactions between people, artefacts, and space' (Pennycook, 2018b, p. 10). The particular, material context in which communication happens is thus not 'just' context but rather part of an interactive whole. This in turn illustrates how meaning is made not just in people's heads but also through their bodies; not just through language but also through the more-than-representational; not just by humans but also by the relations of objects, animals, and other non-human *things*.

While posthumanism as a term has only recently surfaced in sociocultural linguistics, posthumanist perspectives have permeated the field much longer than that (Lamb & Higgins, 2020; see also Bucholtz & Hall, 2016, pp. 187–189). For Lamb and Higgins (2020), this is most evident in Scollon and Scollon's (2004) *nexus analysis*, especially with regards to their work on the emerging internet in Alaska in the late 1970s and early 1980s, where Scollon and Scollon examine the connections between social practices, Alaskan peoples, and technologies. Relatedly, some scholars researching digitally mediated interaction have provided rich analyses of the relationship between language, technology, and the human body (Keating, 2005; Jones, 2009, 2011; Keating & Sunakawa, 2010; Licoppe & Morel, 2012). Elsewhere, scholars in digital discourse studies have also been exploring the role of non-human actors (Eisenlauer, 2014; Procházka, 2019; O'Halloran, 2022; see, especially, also Jones, 2020b, 2021, on 'algorithmic pragmatics'). More broadly, sociocultural linguists have paid increasing attention to matters such as human–animal communication (e.g. Lamb, 2020; Cornips & van den Hengel, 2021); human–machine interactions in non-digital contexts (e.g. Hovens, 2022 on machine-actants in a metal foundry); and sensory, non-material aspects of communication (e.g. Thurlow, 2016; see also Pennycook & Otsuji, 2015, on smell). All in all, posthumanism productively challenges established sociolinguistic understandings of language, agency, materiality, and practice.

Against this theoretical backdrop, I see a posthumanist sociolinguistics of digital media entailing at least three interconnected dimensions or interventions:

1. First, a posthumanist sociolinguistics of digital media centres the dynamic and emergent nature of the interface, foregrounding embodied experience and usage instead of (only) linguistic and/or textual aspects.
2. Second, and in line with the recent emergence of postdigital perspectives (e.g. Jandric et al., 2018), it interrogates these experience(s) by resurfacing and denaturalizing the digital.
3. Third, and overall, it addresses how the notion of 'human' – and who is considered human – surfaces in people's interactions with digital media as well as how 'human' is purposefully constructed in the production of digital media.

These aspects are necessarily interconnected; I therefore discuss them not separately but as an overall posthumanist approach, which starts with acknowledging that digital media are ubiquitous in most people's lives today: they surface in even the most banal, everyday encounters. It is because of this that I propose a critical re-engagement with the digital, one that focuses less on the uses of digital media (such as how people's communicative behaviour 'online' might differ from what they do 'offline') and instead engages the production of digital media and the way digital media are central to shaping the kinds of interactions that (can) take place. As said, this aligns nicely with the recent turn to the postdigital in much of the humanities (e.g. Jandrić & Knox, 2021; Tagg & Lyons, 2022; cf. also Mathier, 2024); here, scholars note how digital media are no longer 'new' and that they can no longer be seen as separate from ordinary human experience (cf. Jandrić et al., 2018) – that, perhaps, the digital and the human have never been separate.

Like posthumanism, the term postdigital is an umbrella term that has been applied to a variety of phenomena across a variety of disciplines (Taffel,

2016). Much of this scholarship is inspired by Negroponte's 1998 article in the technology and business magazine *Wired* titled 'Beyond Digital', in which he declares: 'Face it – the digital revolution is over' (cf. Jandrić et al., 2022). In this way, the prefix 'post' signals both a rupture and a continuation; it is understood as signalling not only a move beyond the digital as new and novel but also as a way of rejecting a dualistic logic of 'on/off' that understands the digital as discrete and in opposition to the continuous nature of human biological existence (Jandrić et al., 2018; cf. also Pepperell & Punt, 2000). As such, the postdigital also draws attention to sociomaterial aspects of digital media, such as the increasingly corporatized and platformized nature of their production (e.g. Knox, 2019; Mathier, 2024), recognizing, as I have in this book, that the production of digital media is neither neutral nor apolitical. Here, too, the digital is understood as something that is omnipresent but which still (or all the more) needs to be interrogated.

Not unlike the notion of posthumanism itself, the term postdigital, as it has been used by scholars in the humanities, is thus particularly useful for highlighting the 'subtle cultural shifts' (Cramer, 2015, p. 15) in how the digital is perceived today: as something that is no longer extraordinary, but simply part of everyday life. This is certainly not news to scholars in digital discourse studies, who have long recognized that online and offline practices are always inherently connected (e.g. Jones, 2004; Georgakopoulou, 2006; for a recent discussion, see Bolander & Locher, 2020). My point, however, is that now that the digital is often taken for granted (by scholars and users alike), it is necessary to resurface and denaturalize digital media, acknowledging that they are never just 'there', but rather purposefully designed. For example, it may be banal and unspectacular that I can read and reply to my work emails from my phone, but I can only do so because someone decided that this is what I *should* be able to do. In talking about the postdigital, I am therefore less interested in aspects of corporatization or platformization (though this certainly matters as well) and more so in a critical reengagement of digital media as something that is actively constructed by particular people and corporations with particular, often self-interested goals.

An approach that centres the constructedness of digital media inevitably needs to investigate the design choices of producers and their implications for users. This, I suggest, is most fruitfully done by focusing not on textual representation but on interaction and embodied experience. In this regard, scholars in media and communication studies (e.g. Light et al., 2018; Dieter & Tkacz, 2020) have proposed walkthrough methods as a way of studying software, an approach which invites researchers to centre the medium and its design by employing a step-by-step walkthrough technique during which researchers systematically click through software interfaces and record interface actions order to understand how the software guides users. Approaches like this also lend themselves well to decentring linguistic-representational aspects (see Ash et al., 2017). Centrally, walkthrough methods allow for a meticulous *surfacing* of the software interface, whereby its affordances and constraints are systematically examined. This, in turn, can reveal what kinds of ideal (or idealized) users are embedded in the software (cf. Chapter 6), all of which is connected to how the category of 'human' is conceptualized and designed in digital media. Just as the humanism of the Enlightenment grants the status of 'human' only to some people, idealized users, too, can reveal something about who is considered more and who is considered less human.

More generally, looking for the non- or posthuman in digital media may seem obvious. Especially in times of artificial intelligence and virtual agents, it is certainly tempting to search for the posthuman in interactions people might have with smart assistants like Siri or Alexa. While encounters with virtual agents like these are interesting to consider, my primary concern is, once again, not the use of digital media but their production, and, specifically, their production by people like UX writers. In this regard, the posthuman aspects of digital media that UX writers surface are not human–machine interactions but rather the more subtle ways in which questions of the human, and of who is considered 'more' or 'less' human, surface in the production of digital media. In fact, one need not look far for this. With regards to virtual agents, Sweeney (2016) shows that anthropomorphized software interfaces are frequently riddled with stereotypical and problematic assumptions about gender and race; this, in turn is connected to how the base category of 'human' is considered in the design of these virtual agents. These issues have long been a concern for race and media scholars like Kolko and Nakamura (e.g. Kolko et al., 2000; Nakamura, 2008; Nakamura & Chow-White, 2012). The problem, suggests Sweeney (2016), is that designers tend to treat believability and humanness as natural categories without questioning their history and social construction. Designing a white virtual agent can then be framed merely as a matter of choosing a face that is – supposedly – more 'believable', 'trustworthy', and 'human'. As a result, racism and sexism too often remain unacknowledged, simply being dismissed as part of (a particular) human nature.

With Siri, Cortana, and Alexa, virtual agents have become almost omnipresent. As such, the feminization of virtual agents and the ensuing feminization of communicative labour (cf. Cameron, 2000b) has received not only scholarly but also public attention. To give but two examples, in May 2019, the *New York Times* reported that 'Siri and Alexa reinforce gender bias'; around the same time, the American business magazine *Fast Company* published an article titled 'Siri, Cortana, Alexa, and others are setting women back in the workplace', which reported on the problematic ways in which voice assistants reinforce the ideological belief that women are supposed to listen and serve.[5] Not surprisingly, the question of race or the fact that these virtual agents tend to have a supposedly ‘neutral’ accent have received less public attention. Again, this is often simply accepted – also by users – as natural and/or normal.

Next to the obvious example of virtual agents, the question of who is considered ‘more’ or ‘less’ human – who is supposed to command and who is supposed to serve – also surfaces in more banal and less obvious ways. A case in point is the automatic spellchecking most people rely on nowadays. Such software automatically corrects misspellings, capitalizes words, or even points out grammar mistakes. Again, from a posthumanist viewpoint, it is certainly interesting to consider where agency lies in this instance: here, humans and machines do not just interact but co-act. But more interesting to me are the issues that surface when one turns to the production of this software and how the decisions of UX writers, designers, or engineers are reflected in what users can in fact do with the software (or not). Once again, this is a place where the political nature of software and algorithms surfaces.

Deciding what is considered ‘correct’ spelling is, as language scholars know only too well, an ideological matter of prescription; decisions on linguistic correctness are often connected to social stigmatization and discrimination (e.g. Milroy & Milroy, 1999; Sebba, 2007; Cameron, 2012). Effectively, then, spellcheck algorithms are a digital form of what Cameron (2012) calls ‘verbal hygiene’, an intentional attempt to correct and control language in order to make it conform to a particular norm. Importantly, this happens less through direct representation and more so through people’s interactions with the software, and it is again entangled with specific understandings of who is considered (more) human.

As science and technology scholar Benjamin (2019) points out, autocorrections can indicate which phenomena are deemed legitimate and which are considered a mistake or a myth. Spellcheck algorithms, Benjamin observes, used to readily accept the word ‘underserved’ as correct but mark the word ‘overserved’ as something that is made up. As she points out, a word signalling undue privilege thus becomes illegible through the software’s algorithm and interface. Similarly, non-Western names are still frequently underlined in red, whereas Western names are usually accepted as ‘correct’ by spellcheck algorithms. Even if banal, these small moments can add up as technological microaggressions that tell people that they don’t belong. Built by humans

[5] Sources: Siri and Alexa Reinforce Gender Bias, U.N. Finds. Retrieved from www.nytimes.com/2019/05/22/world/siri-alexa-ai-gender-bias.html (accessed 10 October 2022); Why Siri, Cortana, Alexa, and others are setting women back in the workplace. Retrieved from www.fastcompany.com/90426299/why-siri-cortana-alexa-and-others-are-setting-women-back-in-the-workplace (accessed 10 October 2022).

but scaled through technology, interfaces can thus relentlessly re-establish existing inequalities. Importantly, this occurs not so much through the individual interface text but through the repetitive and cumulative usage and embodied experience of these texts.

Ultimately, interfaces exert their power not just through representation but also – or perhaps especially – through non- or more-than-representational aspects. As a result, fully understanding the cultural-political impact of software interfaces necessarily demands a posthumanist approach. I have suggested that denaturalizing these everyday interface practices requires attending also to the production of digital media and to how particular interface practices may be deliberately designed. This in turn can reveal who is considered 'more' human – or the right kind of human – in particular software interfaces. It is in this way that the (post)human is a central component of every software interface.

## 7.5 Concluding Remarks: Language and the Politics of the Interface

Throughout this book, I have argued that interface designs and interface texts significantly shape what people can do and who they can be when using digital media. Interfaces encourage specific performances and therefore constitute a kind of 'action before action' (Friz & Gehl, 2016; cf. also Latour, 2005) that pre-scribes and thus co-constitutes the action of users. The interfaces of digital media may compel users to accept and validate hegemonic discourses through their own actions with and through them – or, to put it more bluntly, interfaces may position everyday users as 'a mechanistic "thing" with faux agency' (Retzepi, 2019, p. 86) rather than as an agent with actual power. I may not want to reinforce gendered views of beauty, but if I want to share with my Facebook friends that I am 'feeling beautiful', I might, for the sake of convenience, do so anyway. I may not want to perform a gendered identity on Pinterest, but Pinterest's tutorials may entice me to do so anyway. I may not want to subscribe to the Western and white views perpetuated through so many contemporary interfaces, but by using digital media, I may end up accepting them as 'normal' anyway.

It is by centring the dynamic nature of interfaces and users' lived, embodied experiences of them that scholars can more fully grasp these cultural politics of software interfaces. This is not to suggest that language does not matter, but it has to be understood as always also connected to the equally important and equally consequential non-representational and embodied aspects of software interfaces. It is through the fleeting, banal, and everyday moments that interfaces most covertly – and maybe most consequentially – exert their power.

In the end, digital media interfaces are perhaps best understood as sites of power, where meaning and culture are negotiated. Addressing this necessarily entails focusing not just on the users of digital media but also on their producers and on resurfacing the user interface as something that is banal and ubiquitous, but also purposefully designed. The way in which these interfaces are written and designed is an inherently normative process by which people like UX writers – consciously or not – project their worldviews onto others, and these worldviews in turn get amplified through people's everyday use of digital media. Here lie the politics and the power of interfaces: they congeal the normative beliefs of those who create and control them. They act as 'entities that don't sleep', ultimately constituting 'associations that don't break down [and] that allow power to last longer and expand further' (Latour, 2005, p. 70). Every time the interface is used, normative beliefs are (re)activated, and it is in this way that the interface is ultimately not just a text but also lived experience by which producers, users, and technology together assert and reproduce normative discourses.

# Appendices

## Appendix 1: Transcription Conventions

I have opted to use simplified transcripts that follow standard orthographic rules in an effort to assist readability. This means that punctuation marks are not meant to indicate intonation patterns but rather that an utterance was interpreted as, for instance, a question; additionally, I have omitted micropauses and more detailed paralinguistic cues (e.g. lengthening of sounds) for ease of reading – except where I discuss such detailed interactional nuances in my analysis.

| | |
|---|---|
| .,? | punctuation marks indicate how an utterance is interpreted |
| – | indicates an abrupt cut-off |
| (.) | indicates a short pause |
| = | indicates latching |
| : | indicates that the previous sound is lengthened |
| >word< | indicates talk that is sped up vis-à-vis surrounding talk |
| (word) | indicates that the exact utterance is unclear from the recording |
| ((word)) | indicates non-verbal communication such as laughter or gestures |
| [word] | indicates overlapping talk |
| *word* | indicates stress |
| . . . | indicates omissions |

## Appendix 2: Primary Sources (Chapters 2 and 3)

Kubie, S. (2018). *Writing for designers*. A Book Apart.

Metts, M. J., & Welfle, A. (2020). *Writing is designing: Words and the user experience*. Rosenfeld Media.

Podmajersky, T. (2019). *Strategic writing for UX: Drive engagement, conversion, and retention with every word.* O'Reilly.

# References

Abbott, A. (1988). *The system of professions: An essay on the division of expert labor*. University of Chicago Press.

Adams, T. L. (2010). Profession: A useful concept for sociological analysis? *Canadian Review of Sociology/Revue canadienne de sociologie*, *47*(1), 49–70. https://doi.org/10.1111/j.1755-618X.2010.01222.x

(2018). Theorizing professions. In *Regulating professions: The emergence of professional self-regulation in four Canadian provinces* (pp. 15–36). University of Toronto Press.

Adams-Thies, B. (2012). Fluid bodies or bodily fluids: Bodily reconfigurations in cybersex. *Journal of Language and Sexuality*, *1*(2), 179–205. https://doi.org/10.1075/jls.1.2.03ada

Agha, A. (1999). Register. *Journal of Linguistic Anthropology*, *9*(1/2), 216–219.

Allan, K. (2013). Skilling the self: The communicability of immigrants as flexible labour. In A. Duchêne, M. Moyer, & C. Roberts (Eds.), *Language, migration and social inequalities* (pp. 55–81). Multilingual Matters. https://doi.org/10.21832/9781783091010-004

Althusser, L. (1971). Ideology and ideological state apparatuses. In *Lenin and philosophy and other essays* (pp. 127–186). Monthly Review Press.

Anderson, B., & Borges-Rey, E. (2019). Encoding the UX: User interface as a site of encounter between data journalists and their constructed audiences. *Digital Journalism*, *7*(9), 1253–1269. https://doi.org/10.1080/21670811.2019.1607520

Androutsopoulos, J. (2014). Languaging when contexts collapse: Audience design in social networking. *Discourse, Context & Media*, *4–5*, 62–73. https://doi.org/10.1016/j.dcm.2014.08.006

Arrivé, M. (2020). Visibilizing invisibility, an introduction. *InMedia*, *8*(2), 1–11. https://doi.org/10.4000/inmedia.2683

Ash, J., Anderson, B., Gordon, R., & Langley, P. (2017). Unit, vibration, tone: A post-phenomenological method for researching digital interfaces. *Cultural Geographies 25(1)*. https://doi.org/10.1177/1474474017726556

Atkins, S., & Carter, R. (2011). Creativity in speech. In J. P. Gee & M. Handford (Eds.), *The Routledge handbook of discourse analysis* (pp. 315–325). Routledge.

Austin, J. L. (1975). *How to do things with words*. Clarendon Press.

Ayaß, R. (2014). Using media as involvement shields. *Journal of Pragmatics*, *72*, 5–17. https://doi.org/10.1016/j.pragma.2014.02.003

Bakhtin, M. M. (1986). *Speech genres and other late essays* (M. Holquist & C. Emerson, Eds.; V. W. McGee, Trans.). University of Texas Press.

Barad, K. M. (2007). *Meeting the universe halfway: Quantum physics and the entanglement of matter and meaning*. Duke University Press.

Baron, J. N., & Bielby, W. T. (1986). The proliferation of job titles in organizations. *Administrative Science Quarterly*, *31*(4), 561–586.

Bauer, J. M., Bergstrøm, R., & Foss-Madsen, R. (2021). Are you sure, you want a cookie? – The effects of choice architecture on users' decisions about sharing private online data. *Computers in Human Behavior, 120*. https://doi.org/10.1016/j.chb.2021.106729

Bell, A. (1984). Language style as audience design. *Language in Society*, *13*(2), 145–204. https://doi.org/10.1017/S004740450001037X

Benjamin, R. (2019). *Race after technology: Abolitionist tools for the new Jim code*. Polity.

Bermejo Fernandez, C., Chatzopoulos, D., Papadopoulos, D., & Hui, P. (2021). This website uses nudging: MTurk workers' behaviour on cookie consent notices. *Proceedings of the ACM on Human-Computer Interaction, 5(CSCW2)*, 1–22. https://doi.org/10.1145/3476087

Bettivia, R. S. (2016). Seeing the unseen: The politics and privileges of invisibility and opacity in the digital archive. In N. Rodríguez Ortega, F. Díez-Platas, & S. Kuivakari (Eds.), *Airing the past: Inquiries into digital memories* (pp. 1–11). Brill. https://doi.org/10.1163/9781848884953_002

Bezemer, J., & Kress, G. (2008). Writing in multimodal texts: A social semiotic account of designs for learning. *Written Communication*, *25*(2), 166–195. https://doi.org/10.1177/0741088307313177

Bhatia, V. K. (2012). Creative exploitation of socio-pragmatic space in professional discourse. In R. H. Jones (Ed.), *Discourse and creativity* (pp. 75–91). Pearson.

(2016). Creativity and interdiscursive performance in professional communication. In R. H. Jones (Ed.), *The Routledge handbook of language and creativity* (pp. 158–169). Routledge.

Bivens, R. (2017). The gender binary will not be deprogrammed: Ten years of coding gender on Facebook. *New Media & Society*, *19*(6), 880–898. https://doi.org/10.1177/1461444815621527

Block, D. (2017). Political economy in applied linguistics research. *Language Teaching*, *50*(1), 32–64. https://doi.org/10.1017/S0261444816000288

(2018). What on earth is 'language commodification'? In B. Schmenk, S. Breidbach, & L. Küster (Eds.), *Sloganization in language education discourse* (pp. 121–141). Multilingual Matters. https://doi.org/10.21832/9781788921879-008

(2021). Coda: Issues arising around conceptual and empirical work on the commodification of language. In J. E. Petrovic & B. Yazan (Eds.), *The commodification of language: Conceptual concerns and empirical manifestations* (pp. 200–211). Routledge.

Blommaert, J. (2013). Writing as a sociolinguistic object. *Journal of Sociolinguistics*, *17*(4), 440–459. https://doi.org/10.1111/josl.12042

Blommaert, J., & Jie, D. (2010). *Ethnographic fieldwork: A beginner's guide*. Multilingual Matters.

Blommaert, J., & Rampton, B. (2011). Language and superdiversity. *Diversities*, *13*(2), 1–21.

Boden, M. A. (2004). *The creative mind: Myths and mechanisms* (2nd ed). Routledge.

Bolander, B., & Locher, M. A. (2020). Beyond the online offline distinction: Entry points to digital discourse. *Discourse, Context & Media*, *35*, 1–8. https://doi.org/10.1016/j.dcm.2020.100383

Bolchini, D. (2015). User experience and usability. In L. Cantoni & J. A. Danowski (Eds.), *Communication and technology* (pp. 233–248). De Gruyter. https://doi.org/10.1515/9783110271355-014

Bollini, L. (2017). Beautiful interfaces: From user experience to user interface design. *The Design Journal*, *20*(sup1), S89–S101. https://doi.org/10.1080/14606925.2017.1352649

Bolter, J. D., & Grusin, R. A. (1999). *Remediation: Understanding new media*. MIT Press.

Borning, A., Friedman, B., & Logler, N. (2020). The 'invisible' materiality of information technology. *Communications of the ACM*, *63*(6), 57–64. https://doi.org/10.1145/3360647

Bourdieu, P. (1977). The economics of linguistic exchanges. *Social Science Information*, *16*(6), 645–668.

(1986). The forms of capital. In J. E. Richardson (Ed.), *Handbook of theory and research for the sociology of education* (pp. 241–258). Greenwood Press.

(1990). *The logic of practice*. Stanford University Press.

(1991). *Language and symbolic power* (J. B. Thompson, Ed.; G. Raymond & M. Adamson, Trans.). Polity Press.

(1993). *The field of cultural production: Essays on art and literature*. Columbia University Press.

Bourdieu, P., & Wacquant, L. J. D. (1989). Towards a reflexive sociology: A workshop with Pierre Bourdieu. *Sociological Theory*, *7*(1), 26–63.

(1992). *An invitation to reflexive sociology*. University of Chicago Press.

Boutet, J. (2012). Language workers: Emblematic figures of late capitalism. In A. Duchêne & M. Heller (Eds.), *Language in late capitalism: Pride and profit* (pp. 207–229). Routledge.

Boutet, J., & Heller, M. (2007). Enjeux sociaux de la sociolinguistique: Pour une sociolinguistique critique. *Langage et société*, *121–122*(3–4), 305–318.

Boyd, M. S. (2014). (New) participatory framework on YouTube? Commenter interaction in US political speeches. *Journal of Pragmatics*, *72*, 46–58. https://doi.org/10.1016/j.pragma.2014.03.002

Bucher, T., & Helmond, A. (2018). The affordances of social media platforms. In J. Burgess, A. Marwick, & T. Poell (Eds.), *The Sage handbook of social media* (pp. 233–253). Sage.

Bucholtz, M., & Hall, K. (2016). Embodied sociolinguistics. In N. Coupland (Ed.), *Sociolinguistics: Theoretical debates* (pp. 173–197). Cambridge University Press.

Busch, B. (2006). Language biographies: Approaches to multilingualism in education and linguistic research. In B. Busch, A. Jardina, & A. Tjoutuku (Eds.), *Language biographies for multilingual learning* (pp. 5–18). PRAESA Occasional Papers.

(2012). The linguistic repertoire revisited. *Applied Linguistics*, *33*(5), 503–523. https://doi.org/10.1093/applin/ams056

Callon, M., & Latour, B. (1981). Unscrewing the big leviathan: How actors macro-structure reality and how sociologists help them to do so. In K. Knorr Cetina & A. V. Cicourel (Eds.), *Advances in social theory and methodology: Toward an integration of micro and macro sociologies* (pp. 277–303). Routledge.

Cameron, D. (2000a). *Good to talk? Living and working in a communication culture*. Sage.

(2000b). Styling the worker: Gender and the commodification of language in the globalized service economy. *Journal of Sociolinguistics*, *4*(3), 323–347. https://doi.org/10.1111/1467-9481.00119

(2012). *Verbal hygiene*. Routledge. https://doi.org/10.4324/9780203123898

Candlin, C. N., & Sarangi, S. (2003). Introduction: Trading between reflexivity and relevance: New challenges for applied linguistics. *Applied Linguistics*, *24*(3), 271–285.

Candlin, S. (2003). Issues arising when the professional workplace is the site of applied linguistic research. *Applied Linguistics*, *24*(3), 386–394.

Carmi, E. (2021). A feminist Critique to digital consent. *Seminar.Net*, *17*(2). https://doi.org/10.7577/seminar.4291

Carnegie, T. A. M. (2009). Interface as exordium: The rhetoric of interactivity. *Computers and Composition*, *26*(3), 164–173. https://doi.org/10.1016/j.compcom.2009.05.005

Carter, R. (2004). *Language and creativity: The art of common talk*. Routledge.

Carter, R., & McCarthy, M. (2004). Talking, creating: Interactional language, creativity, and context. *Applied Linguistics*, *25*(1), 62–88. https://doi.org/10.1093/applin/25.1.62

Catenaccio, P., Cotter, C., De Smedt, M., Garzone, G., Jacobs, G., Macgilchrist, F., Lams, L., Perrin, D., Richardson, J. E., van Hout, T., & van Praet, E. (2011). Towards a linguistics of news production. *Journal of Pragmatics*, *43*(7), 1843–1852. https://doi.org/10.1016/j.pragma.2010.09.022

Cazden, C. B. (1970). Children's questions: Their forms, functions and roles in education. *Young Children*, *25*(4), 202–220.

Cekaite, A., & Mondada, L. (Eds.). (2020). *Touch in social interaction: Touch, language and body*. Routledge.

Chik, A. (2015). Recreational language learning and digital practices: Positioning and repositioning. In R. H. Jones, A. Chik, & C. A. Hafner (Eds.), *Discourse and digital practices: Doing discourse analysis in the digital age* (pp. 112–129). Routledge.

Chun, W. H. K. (2011). *Programmed visions: Software and memory*. MIT Press.

Clark, D. (2016). Content strategy: An integrative literature review. *IEEE Transactions on Professional Communication*, *59*(1), 7–23. https://doi.org/10.1109/TPC.2016.2537080

Cornips, L., & van den Hengel, L. (2021). Place-making by cows in an intensive dairy farm: A sociolinguistic approach to nonhuman animal agency. In B. Bovenkerk & J. Keulartz (Eds.), *Animals in our midst: The challenges of co-existing with animals in the anthropocene* (pp. 177–201). Springer. https://doi.org/10.1007/978-3-030-63523-7_11

Coupland, N. (2001). Dialect stylization in radio talk. *Language in Society*, *30*(3), 345–375. https://doi.org/10.1017/S0047404501003013

Cramer, F. (2015). What is 'post-digital'? In D. M. Berry & M. Dieter (Eds.), *Postdigital aesthetics* (pp. 12–26). Palgrave Macmillan.

Criado Perez, C. (2019). *Invisible women: Exposing data bias in a world designed for men*. Chatto & Windus.

Cuskley, C., Simner, J., & Kirby, S. (2017). Phonological and orthographic influences in the bouba–kiki effect. *Psychological Research*, *81*(1), 119–130. https://doi.org/10.1007/s00426-015-0709-2

Del Percio, A., Flubacher, M., & Duchêne, A. (2017). Language and political economy. In O. García, N. Flores, & M. Spotti (Eds.), *The Oxford handbook of language and society* (pp. 55–75). Oxford University Press.

Dieter, M. (2015). Dark patterns: Interface design, augmentation and crisis. In D. M. Berry & M. Dieter (Eds.), *Postdigital aesthetics: Art, computation and design* (pp. 163–178). Palgrave Macmillan.

Dieter, M., & Tkacz, N. (2020). The patterning of finance/security: A designerly walkthrough of challenger banking apps. *Computational Culture*, 7. http://computationalculture.net/the-patterning-of-finance-security/

Djonov, E., & van Leeuwen, T. (2013). Between the grid and composition: Layout in PowerPoint's design and use. *Semiotica*, *2013*(197), 1–34. https://doi.org/10.1515/sem-2013-0078

(2017). The power of semiotic software: A critical multimodal perspective. In J. Flowerdew & J. E. Richardson (Eds.), *The Routledge handbook of critical discourse studies* (pp. 566–581). Routledge. https://doi.org/10.4324/9781315739342-39

Dorschel, R. (2022). Reconsidering digital labour: Bringing tech workers into the debate. *New Technology, Work and Employment*, 37(2), 288–307. https://doi.org/10.1111/ntwe.12225

Draucker, F. (2015). Participation structures in Twitter interaction. In M. Dynel & J. Chovanec (Eds.), *Participation in public and social media interactions* (pp. 49–66). John Benjamins.

Droz-dit-Busset, O. (forthcoming). *Disrupting the economies of language work? Social media influencers as 'new generation' copywriters*. Routledge.

Duchêne, A. (2011). Neoliberalism, social inequalities, and multilingualism: The exploitation of linguistic resources and speakers. *Langage et société*, *136*(2), 81–108.

(2020). Unequal language work(ers) in the business of words. In C. Thurlow (Ed.), *The business of words: Wordsmiths, linguists, and other language workers* (pp. 23–35). Routledge.

Duchêne, A., & Heller, M. (Eds.). (2012). *Language in late capitalism: Pride and profit*. Routledge.

Due, B. L., & Licoppe, C. (2020). Video-mediated interaction (VMI): Introduction to a special issue on the multimodal accomplishment of VMI institutional activities. *Social Interaction. Video-Based Studies of Human Sociality*, *3*(3). https://doi.org/10.7146/si.v3i3.123836

Eckert, P. (1996). Vowels and nail polish: The emergence of linguistic style in the preadolescent heterosexual marketplace. In N. Warner, J. Ahlers, L. Bilmes, M. Oliver, S. Wertheim, & M. Chen (Eds.), *Gender and belief systems: Proceedings of the fourth Berkeley women and language conference* (pp. 183–190). Berkeley Women and Language Group.

Ehrensberger-Dow, M., Heeb, A. H., Jud, P., & Angelone, E. (2017). Insights from translation process research in the workplace. In D. Perrin & U. Kleinberger (Eds.), *Doing applied linguistics: Enabling transdisciplinary communication* (pp. 116–123). De Gruyter. https://doi.org/10.1515/9783110496604-014

Ehrlich, S. (2019). 'Well, I saw the picture': Semiotic ideologies and the unsettling of normative conceptions of female sexuality in the Steubenville rape trial. *Gender and Language*, *13*(2), 251–269. https://doi.org/10.1558/genl.35019

Eisenlauer, V. (2014). Facebook as a third author – (Semi-)automated participation framework in Social Network Sites. *Journal of Pragmatics*, *72*, 73–85. https://doi.org/10.1016/j.pragma.2014.02.006

Fairclough, N. (1999). Global capitalism and critical awareness of language. *Language Awareness*, *8*(2), 71–83. https://doi.org/10.1080/09658419908667119

Fishkin, K. P., Moran, T. P., & Harrison, B. L. (1999). Embodied user interfaces: Towards invisible user interfaces. In S. Chatty & P. Dewan (Eds.), *Engineering for human-computer interaction* (pp. 1–18). Springer. https://doi.org/10.1007/978-0-387-35349-4_1

Florida, R. (2012). *The rise of the creative class – revisited: 10th anniversary edition*. Basic Books.

Flubacher, M.-C., Coray, R., & Duchêne, A. (2016). Language, integration, and investment: The regulation of diversity in the context of unemployment. *Multilingua*, *35*(6), 675–696. https://doi.org/10.1515/multi-2015-0075

Foord, J. (2009). Strategies for creative industries: An international review. *Creative Industries Journal*, *1*(2), 91–113. https://doi.org/10.1386/cij.1.2.91_1

Foucault, M. (1972). *The archeology of knowledge: And the discourse on language* (A. M. Sheridan Smith, Trans.). Random House.

Fournier, V. (1999). The appeal to 'professionalism' as a disciplinary mechanism. *The Sociological Review*, *47*(2), 280–307. https://doi.org/10.1111/1467-954X.00173

Frame, A., & Brachotte, G. (2015). *Le tweet stratégique*: Use of Twitter as a PR tool by French politicians. *Public Relations Review*, *41*(2), 278–287. https://doi.org/10.1016/j.pubrev.2014.11.005

Freidson, E. (1983). The theory of professions: State of the art. In R. Dingwall & P. Lewis (Eds.), *The sociology of the professions* (pp. 19–37). Palgrave Macmillan.

Friz, A., & Gehl, R. W. (2016). Pinning the feminine user: Gender scripts in Pinterest's sign-up interface. *Media, Culture & Society*, *38*(5), 686–703. https://doi.org/10.1177/0163443715620925

Galloway, A. R. (2012). *The interface effect*. Polity.

Gaver, W. W. (1991). Technology affordances. *CHI'91*, 79–84.

Georgakopoulou, A. (2006). Postscript: Computer-mediated communication in sociolinguistics. *Journal of Sociolinguistics*, *10*(4), 548–557. https://doi.org/10.1111/j.1467-9841.2006.00292.x

Georgakopoulou, A., & Spilioti, T. (Eds.). (2016). *The Routledge handbook of language and digital communication*. Routledge.

Gershon, I. (2010). Media ideologies: An introduction. *Journal of Linguistic Anthropology*, *20*(2), 283–293. https://doi.org/10.1111/j.1548-1395.2010.01070.x

Gibson, J. J. (1979). *The ecological approach to visual perception*. Houghton Mifflin.

Goffman, E. (1981). *Forms of talk*. University of Pennsylvania Press.

Goodwin, C. (2007). Participation, stance and affect in the organization of activities. *Discourse & Society*, *18*(1), 53–73. https://doi.org/10.1177/0957926507069457

Goodwin, M. H., & Alim, H. S. (2010). "Whatever (neck roll, eye roll, teeth suck)": The situated coproduction of social categories and identities through stancetaking and transmodal stylization. *Journal of Linguistic Anthropology*, *20*(1), 179–194. https://doi.org/10.1111/j.1548-1395.2010.01056.x

Gormley, K. (2020). Neoliberalism and the discursive construction of 'creativity'. *Critical Studies in Education*, *61*(3), 313–328. https://doi.org/10.1080/17508487.2018.1459762

Grant, A. M., Berg, J. M., & Cable, D. M. (2014). Job titles as identity badges: How self-reflective titles can reduce emotional exhaustion. *Academy of Management Journal*, *57*(4), 1201–1225. https://doi.org/10.5465/amj.2012.0338

Gray, C. M., Kou, Y., Battles, B., Hoggatt, J., & Toombs, A. L. (2018). The dark (patterns) side of UX design. *Proceedings of the 2018 CHI Conference on Human Factors in Computing Systems*, 1–14. https://doi.org/10.1145/3173574.3174108

Gray, M. L., & Suri, S. (2019). *Ghost work: How to stop Silicon Valley from building a new global underclass*. Houghton Mifflin Harcourt.

Gunnarsson, B.-L. (2009). *Professional discourse*. Continuum.

Gunnarsson, B.-L., Linell, P., & Nordberg, B. (Eds.). (1997). *The construction of professional discourse*. Addison Wesley.

Hafner, C. A. (2015). Co-constructing identity in virtual worlds for children. In R. H. Jones, A. Chik, & C. A. Hafner (Eds.), *Discourse and digital practices: Doing discourse analysis in the digital age* (pp. 97–111). Routledge.

Håkanson, L. (2007). Creating knowledge: The power and logic of articulation. *Industrial and Corporate Change*, *16*(1), 51–88. https://doi.org/10.1093/icc/dtl033

Hall, C., Sarangi, S., & Slembrouck, S. (1999). The legitimation of the client and the profession: Identities and roles in social work discourse. In S. Sarangi & C. Roberts (Eds.), *Talk, work and institutional order: Discourse in medical, mediation and management settings* (pp. 293–322). Mouton de Gruyter.

Hall, K., Borba, R., & Hiramoto, M. (2020). Language and gender. In J. Stanlaw (Ed.), *The international encyclopedia of linguistic anthropology* (pp. 1–22). Wiley. https://doi.org/10.1002/9781118786093.iela0143

Halliday, M. A. K. (1978). *Language as social semiotic: The social interpretation of language and meaning*. University Park Press.

(1985). Appendix 2: A note on the grammar of little texts. In *An introduction to functional grammar* (pp. 372–377). Edward Arnold.

Hanks, W. F. (2005). Pierre Bourdieu and the practices of language. *Annual Review of Anthropology*, *34*, 67–83. https://doi.org/10.1146/annurev.anthro.33.070203.143907

Haraway, D. (1991). *Simians, cyborgs, and women: The reinvention of nature*. Routledge.

Haraway, D., & Goodeve, T. (1998). *How like a leaf: An interview with Thyrza Nichols Goodeve*. Routledge. https://doi.org/10.4324/9781315022888

Hart-Davidson, W., Spinuzzi, C., & Zachry, M. (2006). Visualizing writing activity as knowledge work: Challenges & opportunities. *Proceedings of the 24th Annual Conference on Design of Communication – SIGDOC'06*, 70–77. https://doi.org/10.1145/1166324.1166341

Hassenzahl, M. (2008). User experience (UX): Towards an experiential perspective on product quality. *Proceedings of the 20th Conference on l'Interaction Homme-Machine – IHM'08*, 11–15. https://doi.org/10.1145/1512714.1512717

Hausendorf, H. (2009). Kleine Texte – über Randerscheinungen von Textualität. *Germanistik in Der Schweiz*, *6*, 5–19.

Heller, M. (2003). Globalization, the new economy, and the commodification of language and identity. *Journal of Sociolinguistics*, *7*(4), 473–492. https://doi.org/10.1111/j.1467-9841.2003.00238.x

(2010a). Language as resource in the globalized new economy. In N. Coupland (Ed.), *The handbook of language and globalization* (pp. 349–365). Wiley-Blackwell.

(2010b). The commodification of language. *Annual Review of Anthropology*, *39*(1), 101–114. https://doi.org/10.1146/annurev.anthro.012809.104951

Heller, M., & Duchêne, A. (2016). Treating langauge as an economic resource: Discourse, data and debate. In N. Coupland (Ed.), *Sociolinguistics: Theoretical debates* (pp. 139–156). Cambridge University Press.

Heller, M., Pujolar, J., & Duchêne, A. (2014). Linguistic commodification in tourism. *Journal of Sociolinguistics*, *18*(4), 539–566. https://doi.org/10.1111/josl.12082

Herbrechter, S. (2013). *Posthumanism: A critical analysis*. Bloomsbury.

Herbrechter, S., Callus, I., de Bruin-Molé, M., Grech, M., Müller, C. J., & Rossini, M. (2020). Critical posthumanism: An overview. In S. Herbrechter, I. Callus, M. Rossini, M. Grech, M. de Bruin-Molé, & C. John Müller (Eds.), *Palgrave handbook of critical posthumanism* (pp. 1–24). Springer.

Herring, S., Stein, D., & Virtanen, T. (Eds.). (2013). *Pragmatics of computer-mediated communication*. De Gruyter.

Hicks, T., & Perrin, D. (2014). Beyond single modes and media: Writing as an ongoing multimodal text production. In E.-M. Jakobs & D. Perrin (Eds.), *Handbook of writing and text production* (pp. 231–253). De Gruyter.

Hookway, B. (2014). *Interface*. MIT Press.

Hopkins, J. (2020). The concept of affordances in digital media. In H. Friese, M. Nolden, G. Rebane, & M. Schreiter (Eds.), *Handbuch Soziale Praktiken und Digitale Alltagswelten* (pp. 47–54). Springer. https://doi.org/10.1007/978-3-658-08357-1_67

Hounshell, J., & Todd, E. (2016). The invisible interface: Traditional design principles in modern electronic design. *20th International Conference on Information Visualisation (IV)*, 375–379. https://doi.org/10.1109/IV.2016.32

Hovens, D. (2022). Breakdowns and assemblages: Including machine-actants in sociolinguistic ethnographies of blue-collar work environments. *Journal of Sociolinguistics*, 27(1), 3–23. https://doi.org/10.1111/josl.12565

Hutchby, I. (2001). Technologies, texts and affordances. *Sociology*, *35*(2), 441–456. https://doi.org/10.1177/S0038038501000219

Iedema, R. (2003). Multimodality, resemiotization: Extending the analysis of discourse as multi-semiotic practice. *Visual Communication*, *2*(1), 29–57. https://doi.org/10.1177/1470357203002001751

Irani, L. C., & Silberman, M. S. (2013). Turkopticon: Interrupting worker invisibility in amazon mechanical turk. *Proceedings of the SIGCHI Conference on Human Factors in Computing Systems*, 611–620. https://doi.org/10.1145/2470654.2470742

Irvine, J. T. (1989). When talk isn't cheap: Language and political economy. *American Ethnologist*, *16*(2), 248–267.

Irvine, J. T., & Gal, S. (2000). Language ideology and linguistic differentiation. In P. V. Kroskrity (Ed.), *Regimes of language* (pp. 35–83). School of American Research.

Ivanova, M., Bronowicka, J., Kocher, E., & Degner, A. (2018). Foodora and Deliveroo: The app as a boss? Control and autonomy in app-based management – the case of food delivery riders. *Working Paper Forschungsförderung*, *107*, 1–51.

Jacobs, G. (2020). Unwriteable discourse? Co-crafting the language of science news. In C. Thurlow (Ed.), *The business of words: Wordsmiths, linguists, and other language workers* (pp. 53–66). Routledge.

Jaffe, A. (2020). Framing elite knowledge in shifting linguistic economies: The case of minority language translation. In C. Thurlow (Ed.), *The business of words: Wordsmiths, linguists, and other language workers* (pp. 141–162). Routledge.

Jakobson, R. (1960). Closing statement: Linguistics and poetics. In T. A. Sebeok (Ed.), *Style in Language* (pp. 350–377). MIT Press.

Jandrić, P., & Knox, J. (2021). The postdigital turn: Philosophy, education, research. *Policy Futures in Education*, 20(7), 780–795. https://doi.org/10.1177/14782103211062713

Jandrić, P., Knox, J., Besley, T., Ryberg, T., Suoranta, J., & Hayes, S. (2018). Postdigital science and education. *Educational Philosophy and Theory*, *50*(10), 893–899. https://doi.org/10.1080/00131857.2018.1454000

Jandrić, P., MacKenzie, A., & Knox, J. (2022). Postdigital research: Genealogies, challenges, and future perspectives. *Postdigital Science and Education* *6*(1), n.p.. https://doi.org/10.1007/s42438-022-00306-3

Jaworski, A. (2017). Epilogue: The moiré effect and the art of assemblage. *Social Semiotics*, *27*(4), 532–543. https://doi.org/10.1080/10350330.2017.1334405

Jensen, J. (2013). IT and experiences: User experience, experience design and user-experience design. In J. Sundbo & F. Sørensen (Eds.), *Handbook on the experience economy* (pp. 179–208). Edward Elgar Publishing.

Joerges, B. (1999). Do politics have artefacts? *Social Studies of Science*, *29*(3), 411–431. https://doi.org/10.1177/030631299029003004

John, N. A. (2017). *The age of sharing*. Polity.

Jones, R. H. (2004). The problem of context in computer mediated communication. In P. LeVine & R. Scollon (Eds.), *Discourse and technology: Multimodal discourse analysis* (pp. 20–33). Georgetown University Press.

(2009). Dancing, skating and sex: Action and text in the digital age. *Journal of Applied Linguistics and Professional Practice*, *6*(3), 283–302. https://doi.org/10.1558/japl.v6i3.283

(2010). Creativity and discourse. *World Englishes*, *29*(4), 467–480. https://doi.org/10.1111/j.1467-971X.2010.01675.x

(2011). C me Sk8: Discourse, technology, and "bodies without organs". In C. Thurlow & K. Mroczek (Eds.), *Digital discourse: Language in the new media* (pp. 321–339). Oxford University Press.

(2012). Introduction: Discourse and creativity. In R. H. Jones (Ed.), *Discourse and creativity* (pp. 1–13). Pearson.

(2015). Discourse, cybernetics, and the entextualisation of the self. In R. H. Jones, A. Chik, & C. A. Hafner (Eds.), *Discourse and digital practices: Doing discourse analysis in the digital age* (pp. 28–47). Routledge.

(2016). Creativity and discourse analysis. In R. H. Jones (Ed.), *The Routledge handbook of language and creativity* (pp. 61–77). Routledge.

(2018). Messy creativity. *Language Sciences*, *65*, 82–86.

(2020a). Accounting for surveillance. *Journal of Sociolinguistics*, *24*(1), 89–95. https://doi.org/10.1111/josl.12405

(2020b). The rise of the pragmatic web: Implications for rethinking meaning and interaction. In C. Tagg & M. Evans (Eds.), *Message and medium: English language practices across old and new media* (pp. 17–37). De Gruyter. https://doi.org/10.1515/9783110670837-003

(2020c). Towards an embodied visual semiotics: Negotiating the right to look. In C. Thurlow, C. Dürscheid, & F. Diémoz (Eds.), *Visualizing digital discourse: Interactional, institutional and ideological perspectives* (pp. 19–42). De Gruyter. https://doi.org/10.1515/9781501510113-002

(2021). The text is reading you: Teaching language in the age of the algorithm. *Linguistics and Education*, *62*, 1–7. https://doi.org/10.1016/j.linged.2019.100750

(2022a). Is my mobile phone listening to me? Conspiratorial thinking, digital literacies, and everyday encounters with surveillance. In M. Demata, V. Zorzi, & A. Zottola (Eds.), *Conspiracy theory discourses* (pp. 49–70). John Benjamins.

(2022b). *Synthetic embodiment and metapragmatic entanglements: Mediated discourse beyond binaries*. Sociolinguistics Symposium 24, Ghent, Belgium.

Jones, R. H., Bhatia, V. K., Bremner, S., & Peirson-Smith, A. (2012). Creative collaboration in the public relations industry. In R. H. Jones (Ed.), *Discourse and creativity* (pp. 93–107). Pearson.

Jones, R. H., Chik, A., & Hafner, C. A. (Eds.). (2015). *Discourse and digital practices: Doing discourse analysis in the digital age*. Routledge. https://doi.org/10.4324/9781315726465

Kalin, N. M. (2016). We're all creatives now: Democratized creativity and education. *Journal of the Canadian Association for Curriculum Studies*, *13*(2), 32–44.

Karlander, D. (2017). Roads to regimentation: Place, authenticity and the metapragmatics of naming. *Language & Communication*, *53*, 11–24. https://doi.org/10.1016/j.langcom.2016.11.002

Keane, W. (2003). Semiotics and the social analysis of material things. *Language & Communication*, *23*(3), 409–425. https://doi.org/10.1016/S0271-5309(03)00010-7

(2018). On semiotic ideology. *Signs and Society*, *6*(1), 64–87. https://doi.org/10.1086/695387

Keating, E. (2005). Homo prostheticus: Problematizing the notions of activity and computermediated interaction. *Discourse Studies*, *7*(4–5), 527–545. https://doi.org/10.1177/1461445605054405

Keating, E., & Mirus, G. (2003). American Sign Language in virtual space: Interactions between deaf users of computer-mediated video communication and the impact of technology on language practices. *Language in Society*, *32*(5), 693–714. https://doi.org/10.1017/S0047404503325047

Keating, E., & Sunakawa, C. (2010). Participation cues: Coordinating activity and collaboration in complex online gaming worlds. *Language in Society*, *39*(3), 331–356. https://doi.org/10.1017/S0047404510000217

Kelly-Holmes, H. (2014). Linguistic fetish: The sociolinguistics of visual multilingualism. In D. Machin (Ed.), *Visual communication* (pp. 135–151). De Gruyter.

(2020). The linguistic business of marketing. In C. Thurlow (Ed.), *The business of words: Wordsmiths, linguists, and other language workers* (pp. 36–50). Routledge.

Kendrick, M. (2005). Invisibility, race, and the interface. *Rhetoric Review*, *24*(4), 395–399.

Kirschenbaum, M. G. (2012). *Mechanisms: New media and the forensic imagination*. MIT Press.

Knoblauch, H. (2005). Focused ethnography. *Forum Qualitative Sozialforschung / Forum: Qualitative Social Research*, *6*(3). https://doi.org/10.17169/fqs-6.3.20

Knox, J. (2019). What does the 'postdigital' mean for education? Three critical perspectives on the digital, with implications for educational research and practice. *Postdigital Science and Education*, *1*(2), 357–370. https://doi.org/10.1007/s42438–019-00045-y

Kolko, B. E., Nakamura, L., & Rodman, G. B. (Eds.). (2000). *Race in cyberspace*. Routledge.

Koller, V. (2017). Language awareness and language workers. *Language Awareness*, *27*(1–2), 4–20. https://doi.org/10.1080/09658416.2017.1406491

Kong, K. (2014). *Professional discourse*. Cambridge University Press.

Koskinen, K. (2000). *Beyond ambivalence: Postmodernity and the ethics of translation* [PhD thesis]. University of Tampere.

Kou, Y., & Gray, C. M. (2018). Towards professionalization in an online community of emerging occupation: Discourses among UX practitioners. *Proceedings of the 2018 ACM Conference on Supporting Groupwork*, 322–334. https://doi.org/10.1145/3148330.3148352

Kraft, K. (2020). Trajectory of a language broker: Between privilege and precarity. *International Journal of Multilingualism*, *17*(1), 80–96. https://doi.org/10.1080/14790718.2020.1682256

Kramsch, C. (2009). *The multilingual subject: What foreign language learners say about their experience and why it matters*. Oxford University Press.

(2021). *Language as symbolic power*. Cambridge University Press. https://doi.org/10.1017/9781108869386

Kress, G. (1997). *Before writing: Rethinking the paths to literacy*. Routledge. https://doi.org/10.4324/9780203992692

(2010). *Multimodality: A social semiotic approach to contemporary communication*. Routledge. https://doi.org/10.4324/9780203970034

Kress, G., & van Leeuwen, T. (2001). *Multimodal discourse: The modes and media of contemporary communication*. Arnold.

(2006). *Reading images: The grammar of visual design* (2nd edition). Routledge. (Original work published 1996)

Krug, S. (2014). *Don't make me think, revisited: A common sense approach to Web usability* (3rd edition). New Riders. (Original work published 2000)

Krzyżanowski, M. (2011). Ethnography and critical discourse analysis: Towards a problem-oriented research dialogue. *Critical Discourse Studies*, *8*(4), 231–238. https://doi.org/10.1080/17405904.2011.601630

Kvåle, G. (2018). Expressing identity in Microsoft Word: A critical discussion of the stylistic normativity of templates and software. In C. M. Johannessen & T. van Leeuwen (Eds.), *The materiality of writing: A trace-making perspective* (pp. 157–174). Routledge.

Labov, W. (1972). *Sociolinguistic patterns*. University of Pennsylvania Press.

Lamb, G. (2020). Towards a green applied linguistics: Human–sea turtle semiotic assemblages in Hawai'i. *Applied Linguistics*, *41*(6), 922–946. https://doi.org/10.1093/applin/amz046

Lamb, G., & Higgins, C. (2020). Posthumanism and its implications for discourse studies. In A. De Fina & A. Georgakopoulou (Eds.), *The Cambridge handbook of discourse studies* (1st edition, pp. 350–370). Cambridge University Press.

Landqvist, M. (2019). Semiotic spaces in antidiscriminatory political discourse: Naming practices as indexes. *Language in Society*, *48*(5), 721–743. https://doi.org/10.1017/S0047404519000459

Latour, B. (1984). The powers of association. *The Sociological Review*, *32*(1_suppl), 264–280. https://doi.org/10.1111/j.1467-954X.1984.tb00115.x

(1990). Technology is society made durable. *The Sociological Review*, *38*(1), 103–131. https://doi.org/10.1111/j.1467-954X.1990.tb03350.x

(2005). *Reassembling the social: An introduction to actor-network-theory*. Oxford University Press.

Lazar, M. (2007). Feminist critical discourse analysis: Articulating a feminist discourse praxis. *Critical Discourse Studies*, *4*(2), 141–164. https://doi.org/10.1080/17405900701464816

Lee, E.-J., & Shin, S. Y. (2014). When the medium is the message: How transportability moderates the effects of politicians' Twitter communication. *Communication Research*, *41*(8), 1088–1110. https://doi.org/10.1177/0093650212466407

Lialina, O. (2019). Once again, the doorknob: Affordance, forgiveness, and ambiguity in human-computer interaction and human-robot interaction. *Media Theory*, *3*(1), 49–72.

Licoppe, C., & Morel, J. (2012). Video-in-interaction: "Talking heads" and the multimodal organization of mobile and skype video calls. *Research on Language and Social Interaction*, *45*(4), 399–429. https://doi.org/10.1080/08351813.2012.724996

Liddicoat, A. J. (2011). *An introduction to conversation analysis* (2nd ed.). Continuum.

Liep, J. (Ed.). (2001). Introduction. In *Locating cultural creativity* (pp. 1–13). Pluto Press. https://doi.org/10.2307/j.ctt18fs9q6

Light, B., Burgess, J., & Duguay, S. (2018). The walkthrough method: An approach to the study of apps. *New Media & Society*, *20*(3), 881–900. https://doi.org/10.1177/1461444816675438

Lillis, T. (2013). *The sociolinguistics of writing*. Edinburgh University Press.

Lillis, T., & Maybin, J. (2017). Introduction: The dynamics of textual trajectories in professional and workplace practice. *Text & Talk*, *37*(4), 409–414. https://doi.org/10.1515/text-2017-0017

Livingstone, S. (1999). New media, new audiences? *New Media & Society*, *1*(1), 59–66. https://doi.org/10.1177/1461444899001001010

Lønsmann, D., & Kraft, K. (2017). Language in blue-collar workplaces. In B. Vine (Ed.), *The Routledge handbook of language in the workplace (pp. 138–149).* Routledge.

Lorimer, H. (2005). Cultural geography: The busyness of being 'more-than-representational'. *Progress in Human Geography*, *29*(1), 83–94. https://doi.org/10.1191/0309132505ph531pr

Ma, E., & Birrell, E. (2022). Prospective consent: The effect of framing on cookie consent decisions. *CHI Conference on Human Factors in Computing Systems Extended Abstracts*, 1–6. https://doi.org/10.1145/3491101.3519687

Macgilchrist, F. (2017). *Textbook production: The entangled practices of developing educational media for schools* (Eckert. Dossiers 15). Georg Eckert Institute for International Textbook Research. http://repository.gei.de/bitstream/handle/11428/267/ED15_Macgilchrist_Production.pdf

(2020). Beyond the academic 'but': The pleasures and politics of collaborative language work in the publishing industry. In C. Thurlow (Ed.), *The business of words: Wordsmiths, linguists, and other language workers* (pp. 163–178). Routledge.

Macgilchrist, F., Ahlrichs, J., Mielke, P., & Richtera, R. (2017). Memory practices and colonial discourse: On text trajectories and lines of flight. *Critical Discourse Studies*, *14*(4), 341–361. https://doi.org/10.1080/17405904.2017.1325382

Macgilchrist, F., & van Hout, T. (2011). Ethnographic discourse analysis and social science. *Forum Qualitative Sozialforschung / Forum: Qualitative Social Research*, *12*(1). https://doi.org/10.17169/fqs-12.1.1600

Machuletz, D., & Böhme, R. (2020). Multiple purposes, multiple problems: A user study of consent dialogs after GDPR. *Proceedings on Privacy Enhancing Technologies*, *2020*(2), 481–498. https://doi.org/10.2478/popets-2020-0037

Maier, C. D. (2017). Multimodality, synaesthesia and intersemiotic translation. *HERMES - Journal of Language and Communication in Business*, *55*, 7–9. https://doi.org/10.7146/hjlcb.v0i55.25899

Mapes, G. (2021). *Elite authenticity: Remaking distinction in food discourse*. Oxford University Press.

(2023). The life of a political speech(writer): Metadiscursive text trajectories in high-end language work. *Journal of Linguistic Anthropology* 33(3), 231–381.

(2025). *The language work of speech writers*. Cambridge University Press.

Marwick, A. E. (2013). *Status update: Celebrity, publicity, and branding in the social media age*. Yale University Press.

Marwick, A. E., & boyd, d. (2011). I tweet honestly, I tweet passionately: Twitter users, context collapse, and the imagined audience. *New Media & Society*, *13*(1), 114–133. https://doi.org/10.1177/1461444810365313

Marx, K. (2019). *Das Kapital: Kritik der politischen Ökonomie. Erster Band.* Felix Meiner Verlag. (Original work published 1867)

Maryns, K., & Blommaert, J. (2002). Pretextuality and pretextual gaps: On de/refining linguistic inequality. *Pragmatics*, *12*(1), 11–30. https://doi.org/10.1075/prag.12.1.02mar

Massanari, A. L. (2007). *In context: Information architects, politics, and interdisciplinarity* [PhD thesis]. University of Washington.

(2010). Designing for imaginary friends: Information architecture, personas and the politics of user-centered design. *New Media & Society*, *12*(3), 401–416. https://doi.org/10.1177/1461444809346722

Mathier, M. (2024). *Postdigital disconnects: The discursive formation of technology in education*. Routledge.

Maybin, J., & Swann, J. (2007). Everyday creativity in language: Textuality, contextuality, and critique. *Applied Linguistics*, *28*(4), 497–517. https://doi.org/10.1093/applin/amm036

McCorkle, B. (2017). Resisting the "natural": Rhetorical delivery and the natural user interface. In C. R. Miller & A. R. Kelly (Eds.), *Emerging genres in new media environments* (pp. 99–115). Springer. https://doi.org/10.1007/978-3-319-40295-6_5

McGill, K. (2013). Political economy and language: A review of some recent literature. *Journal of Linguistic Anthropology*, *23*(2), 84–101. https://doi.org/10.1111/jola.12015

Meier, Y., Schäwel, J., & Krämer, N. C. (2020). The shorter the better? Effects of privacy policy length on online privacy decision-making. *Media and Communication*, *8*(2), 291–301. https://doi.org/10.17645/mac.v8i2.2846

Merchant, G. (2015). Apps, adults and young children: Researching digital literacy practices in context. In R. H. Jones, A. Chik, & C. A. Hafner (Eds.), *Discourse and digital practices: Doing discourse analysis in the digital age* (pp. 143–157). Routledge.

Milroy, J., & Milroy, L. (1999). *Authority in language: Investigating standard English (3rd edition)*. Routledge.

Mondada, L. (2011). Understanding as an embodied, situated and sequential achievement in interaction. *Journal of Pragmatics*, *43*(2), 542–552. https://doi.org/10.1016/j.pragma.2010.08.019

Monteiro, M. (2019). *Ruined by design: How designers destroyed the world, and what we can do to fix it*. Mule Design.

Monteiro, S. (2017). *The fabric of interface: Mobile media, design, and gender*. MIT Press.

Morson, G. S. (2006). Addressivity. In K. Brown (Ed.), *Encyclopedia of language & linguistics* (2nd edition, pp. 55–58). Elsevier. https://doi.org/10.1016/B0-08-044854-2/00389-8

Moschini, I., & Sindoni, M. G. (2021). Language as the tip of the iceberg? Shedding a critical light on 'hidden' discourse in digital platforms. *Discourse, Context & Media*, *42*, 1–12. https://doi.org/10.1016/j.dcm.2021.100505

Mosco, V. (1996). *The political economy of communication: Rethinking and renewal*. Sage.

Murphy, K. M. (2012). Transmodality and temporality in design interactions. *Journal of Pragmatics*, *44*(14), 1966–1981. https://doi.org/10.1016/j.pragma.2012.08.013

Myers, G. (1992). Textbooks and the sociology of scientific knowledge. *English for Specific Purposes*, *11*(1), 3–17. https://doi.org/10.1016/0889-4906(92)90003-S

Nakamura, L. (2008). Cyberrace. *PMLA*, *123*(5), 1673–1682. https://doi.org/10.1632/pmla.2008.123.5.1673

Nakamura, L., & Chow-White, P. (Eds.). (2012). *Race after the Internet*. Routledge.

Nash, K. (2001). The 'cultural turn' in social theory: Towards a theory of cultural politics. *Sociology*, *35*(1), 77–92. https://doi.org/10.1017/S0038038501000050

Norman, D. A. (1990). Why interfaces don't work. In B. Laurel (Ed.), *The art of human computer interface design* (pp. 209–219). Addison Wesley.

(1998). *The invisible computer: Why good products can fail, the personal computer is so complex, and information appliances are the solution*. MIT Press.

(2002). *The design of everyday things*. Basic Books. (Original work published 1988)

Norman, D. A., & Nielsen, J. (2021). *The definition of user experience (UX)*. https://www.nngroup.com/articles/definition-user-experience/

Norrick, N. R. (2016). Language play in conversation. In N. Bell (Ed.), *Multiple perspectives on language play* (pp. 11–46). De Gruyter. https://doi.org/10.1515/9781501503993-002

Norris, S. (2004). *Analyzing multimodal interaction: A methodological framework*. Routledge.

(2012). The creation of a community artist in everyday life: Long-duration process and creative actions. In R. H. Jones (Ed.), *Discourse and creativity* (pp. 143–163). Pearson.

Norris, S., & Jones, R. H. (Eds.). (2005). *Discourse in action: Introducing mediated discourse analysis*. Routledge.

Nyström, S. (2009). The dynamics of professional identity formation: Graduates' transitions from higher education to working life. *Vocations and Learning*, *2*(1), 1–18. https://doi.org/10.1007/s12186-008-9014-1

Odell, L., & Goswami, D. (1982). Writing in a non-academic setting. *Research in the Teaching of English*, *16*(3), 201–223.

O'Halloran, K. (2022). Posthumanism and corpus linguistics. In A. O'Keeffe & M. McCarthy (Eds.), *The Routledge handbook of corpus linguistics* (pp. 675–692). Routledge.

Opel, D. S., & Hart-Davidson, W. (2019). The primary care clinic as writing space. *Written Communication*, *36*(3), 348–378. https://doi.org/10.1177/0741088319839968

O'Shaugnessy, J., & O'Shaughnessy, N. (2003). *Persuasion in advertising*. Routledge. https://doi.org/10.4324/9780203299982

Page, T. (2014). Skeuomorphism or flat design: Future directions in mobile device User Interface (UI) design education. *International Journal of Mobile Learning and Organisation*, *8*(2), 130. https://doi.org/10.1504/IJMLO.2014.062350

Pappert, S., & Roth, K. S. (Eds.). (2021). *Kleine Texte*. Peter Lang.

Parkin, F. (1979). *Marxism and class theory: A bourgeois critique*. Tavistock Publications.

Patterson, D., & Costain, S. (2015). The effectiveness of transient user interface components. *User Interfaces*, *162*, 3–10.

Peck, A., & Stroud, C. (2015). Skinscapes. *Linguistic Landscape*, *1*(1–2), 133–151. https://doi.org/10.1075/ll.1.1-2.08pec

Pennycook, A. (2018a). *Posthumanist applied linguistics*. Routledge.

(2018b). Posthumanist applied linguistics. *Applied Linguistics*, *39*(4), 445–461. https://doi.org/10.1093/applin/amw016

Pennycook, A., & Otsuji, E. (2015). Making scents of the landscape. *Linguistic Landscape*, *1*(3), 191–212. https://doi.org/10.1075/ll.1.3.01pen

Pepperell, R., & Punt, M. (2000). *The postdigital membrane: Imagination, technology and desire*. Intellect.

Perrin, D., & Kleinberger, U. (Eds.). (2017). *Doing applied linguistics: Enabling transdisciplinary communication*. De Gruyter. https://doi.org/10.1515/9783110496604

Petrick, E. R. (2020). A historiography of human–computer interaction. *IEEE Annals of the History of Computing*, *42*(4), 8–23. https://doi.org/10.1109/MAHC.2020.3009080

Petrovic, J. E., & Yazan, B. (2021a). Language as instrument, resource, and maybe capital, but not commodity: A Marxian clarification. In J. E. Petrovic & B. Yazan (Eds.), *The commodification of language: Conceptual concerns and empirical manifestations* (pp. 24–40). Routledge.

Petrovic, J. E., & Yazan, B. (Eds.). (2021b). *The commodification of language: Conceptual concerns and empirical manifestations*. Routledge.

Pietikäinen, S. (2021). Powered by assemblage: Language for multiplicity. *International Journal of the Sociology of Language*, *2021*(267–268), 235–240. https://doi.org/10.1515/ijsl-2020-0074

Piller, I. (2001). Identity constructions in multilingual advertising. *Language in Society*, *30*(2), 153–186. https://doi.org/10.1017/S0047404501002019

Pink, S., & Morgan, J. (2013). Short-term ethnography: Intense routes to knowing. *Symbolic Interaction*, *36*(3), 351–361. https://doi.org/10.1002/symb.66

Porcello, T. (2004). Speaking of sound: Language and the professionalization of sound-recording engineers. *Social Studies of Science*, *34*(5), 733–758. https://doi.org/10.1177/0306312704047328

Portmann, L. (2022). Crafting an audience: UX writing, user stylization, and the symbolic violence of little texts. *Discourse, Context & Media*, *48, 1–8*. https://doi.org/10.1016/j.dcm.2022.100622

Postill, J. (2016). Remote ethnography: Studying culture from afar. In L. Hjorth, H. Horst, A. Galloway, & G. Bell (Eds.), *The Routledge companion to digital ethnography (pp. 61–69)*. Routledge.

Poulsen, S. V., Kvåle, G., & van Leeuwen, T. (2018). Special issue: Social media as semiotic technology. *Social Semiotics*, *28*(5), 593–600. https://doi.org/10.1080/10350330.2018.1509815

Preston, D. R. (1996). Whaddayaknow?: The modes of folk linguistic awareness. *Language Awareness*, *5*(1), 40–74. https://doi.org/10.1080/09658416.1996.9959890

Procházka, O. (2019). Making sense of Facebook's content moderation: A posthumanist perspective on communicative competence and internet memes. *Signs and Society*, *7*(3), 362–397. https://doi.org/10.1086/704763

Pucheu, D. (2016). Effacer l'interface: Une trajectoire du design l'interaction homme-machine. *Interfaces Numériques*, *5*(2), 257–276. https://doi.org/10.25965/interfaces-numeriques.3044

Puzey, G., & Kostanski, L. (Eds.). (2016). *Names and naming: People, places, perceptions and power*. Multilingual Matters.

Ramachandran, V. S., & Hubbard, E. M. (2001). Synaesthesia—A window into perception, thought and language. *Journal of Consciousness Studies*, *8*(12), 3–34.

Redish, J. (1995). Adding value as a professional technical communicator. *Technical Communication*, *42*(1), 26–39.

Retzepi, K. (2019). You, the user. *Interface Critique Journal*, *2*, 79–87.

Roberts, S. T. (2019). *Behind the screen: Content moderation in the shadows of social media*. Yale University Press.

Roderick, I. (2016). *Critical discourse studies and technology: A multimodal approach to analysing technoculture*. Bloomsbury Academic.

Ross, A. (1990). Hacking away at the counterculture. *Postmodern Culture*, *1*(1). http://pmc.iath.virginia.edu/text-only/issue.990/ross-1.990 and http://pmc.iath.virginia.edu/text-only/issue.990/ross-2.990

Sacks, H. (1987). On the preferences for agreement and contiguity in sequences in conversation. In G. Button & J. R. E. Lee (Eds.), *Talk and social organisation* (pp. 54–69). Multilingual Matters.

Sarangi, S. (2005). The conditions and consequences of professional discourse studies. *Journal of Applied Linguistics*, *2*(3), 371–394. https://doi.org/10.1558/japl.v2i3.371

Scarlett, A., & Zeilinger, M. (2019). Rethinking affordance. *Media Theory*, *3*(1), 1–48.

Schegloff, E. (1982). Discourse as an interactional achievement: Some uses of uh huh and other things that come between sentences. In D. Tannen (Ed.), *Analyzing discourse: Text and talk* (pp. 71–93). Georgetown University Press.

Schieffelin, B. B., Woolard, K. A., & Kroskrity, P. V. (Eds.). (1998). *Language ideologies: Practice and theory*. Oxford University Press.

Schindler, K., & Wolfe, J. (2014). Beyond single authors: Organizational text production as collaborative writing. In E.-M. Jakobs & D. Perrin (Eds.), *Handbook of writing and text production* (pp. 159–173). De Gruyter.

Schön, D. A. (1993). *The reflective practitioner: How professionals think in action*. Basic Books.

Scollon, R., & Scollon, S. B. K. (2004). *Nexus analysis: Discourse and the emerging Internet*. Routledge.

Sebba, M. (2007). *Spelling and society: The culture and politics of orthography around the world*. Cambridge University Press.

Selfe, C. L., & Selfe, R. J. (1994). The politics of the interface: Power and its exercise in electronic contact zones. *College Composition and Communication*, *45*(4), 480–504.

Shankar, S., & Cavanaugh, J. R. (2012). Language and materiality in global capitalism. *Annual Review of Anthropology*, *41*(1), 355–369. https://doi.org/10.1146/annurev-anthro-092611-145811

Silverstein, M. (1979). Language structure and linguistic ideologies. In P. Clyne, W. F. Hanks, & C. Hofbauer (Eds.), *The elements* (pp. 193–248). Chicago Linguistic Society.

(2003). Indexical order and the dialectics of sociolinguistic life. *Language & Communication*, *23*(3–4), 193–229. https://doi.org/10.1016/S0271-5309(03)00013-2

Simpson, W., & O'Regan, J. P. (2018). Fetishism and the language commodity: A materialist critique. *Language Sciences*, *70*, 155–166. https://doi.org/10.1016/j.langsci.2018.05.009

Singh, A. K., Upadhyaya, N., Seth, A., Hu, X., Sastry, N., & Mondal, M. (2022). What cookie consent notices do users prefer: A study in the wild. *Proceedings of the 2022 European Symposium on Usable Security*, 28–39. https://doi.org/10.1145/3549015.3555675

Sjöström, J., & Goldkuhl, G. (2004). The semiotics of user interfaces: A socio-pragmatic perspective. In K. Liu (Ed.), *Virtual, distributed and flexible organisations: Studies in organisational semiotics* (pp. 217–236). Springer. https://doi.org/10.1007/1-4020-2162-3_14

Slack, J. D., & Wise, J. M. (2014). *Culture and technology: A primer (2nd edition)*. Peter Lang.

Slattery, S. (2007). Undistributing work through writing: How technical writers manage texts in complex information environments. *Technical Communication Quarterly*, *16*(3), 311–325. https://doi.org/10.1080/10572250701291046

Smith, D. E. (2005). *Institutional ethnography: A sociology for people*. AltaMira.

Spitzmüller, J. (2013). *Graphische Variation als soziale Praxis: Eine soziolinguistische Theorie zu skripturaler 'Sichtbarkeit'*. De Gruyter.

(2015). Graphic variation and graphic ideologies: A metapragmatic approach. *Social Semiotics*, *25*(2), 126–141. https://doi.org/10.1080/10350330.2015.1010323

Stanfill, M. (2015). The interface as discourse: The production of norms through web design. *New Media & Society*, *17*(7), 1059–1074. https://doi.org/10.1177/1461444814520873

Star, S. L., & Strauss, A. (1999). Layers of silence, arenas of voice: The ecology of visible and invisible work. *Computer Supported Cooperative Work*, *8*, 9–30.

Steiger, T. L. (1993). Construction skill and skill construction. *Work, Employment and Society*, *7*(4), 535–560.

Stilwell, F. (2019). From economics to political economy: Contradictions, challenge, and change. *American Journal of Economics and Sociology*, *78*(1), 35–62. https://doi.org/10.1111/ajes.12259

Stöckl, H. (2014). Typography. In S. Norris & C. D. Maier (Eds.), *Interactions, images and texts: A reader in multimodality* (pp. 283–295). De Gruyter. https://doi.org/10.1515/9781614511175.281

Susser, D. (2019). Invisible influence: Artificial intelligence and the ethics of adaptive choice architectures. *Proceedings of the 2019 AAAI/ACM Conference on AI, Ethics, and Society*, 403–408. https://doi.org/10.1145/3306618.3314286

Swann, J., & Deumert, A. (2018). Sociolinguistics and language creativity. *Language Sciences*, *65*, 1–8. https://doi.org/10.1016/j.langsci.2017.06.002

Swarts, J. (2010). Recycled writing: Assembling actor networks from reusable content. *Journal of Business and Technical Communication*, *24*(2), 127–163. https://doi.org/10.1177/1050651909353307

Sweeney, M. E. (2016). The intersectional interface. In S. U. Noble & B. M. Tynes (Eds.), *The intersectional internet: Race, sex, class, and culture online* (pp. 215–228). Peter Lang.

Taffel, S. (2016). Perspectives on the postdigital: Beyond rhetorics of progress and novelty. *Convergence*, *22*(3), 324–338. https://doi.org/10.1177/1354856514567827

Tagg, C., & Lyons, A. (2022). *Mobile messaging and resourcefulness: A post-digital ethnography*. Routledge.

Tan, P. K. W. (2021). Names as linguistic capital. In J. E. Petrovic & B. Yazan (Eds.), *The commodification of language* (pp. 164–180). Routledge.

Thrift, N. (2007). *Non-representational theory: Space, politics, affect*. Sage.

Thurlow, C. (2006). From statistical panic to moral panic: The metadiscursive construction and popular exaggeration of new media language in the print media. *Journal of Computer-Mediated Communication*, *11*(3), 667–701. https://doi.org/10.1111/j.1083-6101.2006.00031.x

(2012). Determined creativity: Language play in new media. In R. H. Jones (Ed.), *Discourse and creativity* (pp. 169–210). Pearson.

(2015). Multimodality, materiality and everyday textualities: The sensuous stuff of status. In G. Rippl (Ed.), *Handbook of intermediality: Literature – image – sound – music* (pp. 619–636). De Gruyter. https://doi.org/10.1515/9783110311075-035

(2016). Queering critical discourse studies or/and performing 'post-class' ideologies. *Critical Discourse Studies*, *13*(5), 485–514. https://doi.org/10.1080/17405904.2015.1122646

(2017). "Forget about the words"? Tracking the language, media and semiotic ideologies of digital discourse: The case of sexting. *Discourse, Context & Media*, *20*, 10–19. https://doi.org/10.1016/j.dcm.2017.06.008

(2018). Critical discourse studies in/of applied contexts: Missed opportunities, fraught possibilities. In J. Flowerdew & J. E. Richardson (Eds.), *The Routledge handbook of critical discourse studies* (pp. 328–341). Routledge.

(2019). Semiotic creativities in and with space: Binaries and boundaries, beware! *International Journal of Multilingualism*, *16*(1), 94–104. https://doi.org/10.1080/14790718.2018.1500264

(2020a). Expanding our sociolinguistic horizons? Geographical thinking and the articulatory potential of commodity chain analysis. *Journal of Sociolinguistics*, *24*(3), 350–368. https://doi.org/10.1111/josl.12388

(Ed.). (2020b). *The business of words: Wordsmiths, linguists, and other language workers*. Routledge.

(2020c). The (grubby) business of words: What 'George Clooney' tells us. In C. Thurlow (Ed.), *The business of words: Wordsmiths, linguists, and other language workers* (pp. 1–19). Routledge.

(2020d). Touching taste: The classist intersections of materiality, textuality and tactility. *Multimodal Communication*, *9*(2), 1–16. https://doi.org/10.1515/mc-2020-0001

(2022). Spracharbeit im Filmgeschäft: Von der Bedeutung "kleiner Texte". In S. Brommer, K. S. Roth, & J. Spitzmüller (Eds.), *Brückenschläge: Linguistik an den Schnittstellen. Festschrift für Christa Dürscheid* (pp. 155–175). Narr.

Thurlow, C., & Britain, D. (2020). Voice work: Learning about and from dialect coaches. In C. Thurlow (Ed.), *The business of words: Wordsmiths, linguists, and other language workers* (pp. 67–85). Routledge.

Thurlow, C., & Jaworski, A. (2006). The alchemy of the upwardly mobile: Symbolic capital and the stylization of elites in frequent-flyer programmes. *Discourse & Society*, *17*(1), 99–135. https://doi.org/10.1177/0957926506058066

(2013). Visible-invisible: The social semiotics of labour in luxury tourism. In T. Birtchnell & J. Caletrío (Eds.), *Elite Mobilities* (pp. 176–193). Routledge.

(2017a). Introducing elite discourse: The rhetorics of status, privilege, and power. *Social Semiotics*, *27*(3), 243–254. https://doi.org/10.1080/10350330.2017.1301789

(2017b). Word-things and thing-words: The transmodal production of privilege and status. In J. R. Cavanaugh & S. Shankar (Eds.), *Language and materiality: Ethnographic and theoretical explorations* (pp. 185–203). Cambridge University Press.

Thurlow, C., & Mroczek, K. (Eds.). (2011). *Digital discourse: Language in the new media*. Oxford University Press.

Tognini-Bonelli, E. (1993). Interpretative nodes in discourse: Actual and actually. In M. Baker, G. Francis, & E. Tognini-Bonelli (Eds.), *Text and technology: In honour of John Sinclair*. John Benjamins.

Toolan, M. (2012). Poems: Wonderfully repetitive. In R. H. Jones (Ed.), *Discourse and creativity* (pp. 17–34). Pearson.

Turner, P. (2017). *A psychology of user experience*. Springer. https://doi.org/10.1007/978-3-319-70653-5

Urciuoli, B. (2008). Skills and selves in the new workplace. *American Ethnologist*, *35*(2), 211–228. https://doi.org/10.1111/j.1548-1425.2008.00031.x

(2016). Discussion: Imagining the languaged worker's language. *Multilingua*, *35*(4), 469–481. https://doi.org/10.1515/multi-2015-0016

Urciuoli, B., & LaDousa, C. (2013). Language management/labor. *Annual Review of Anthropology*, *42*(1), 175–190. https://doi.org/10.1146/annurev-anthro-092412-155524

Utz, C., Degeling, M., Fahl, S., Schaub, F., & Holz, T. (2019). (Un)informed consent: Studying GDPR consent notices in the field. *Proceedings of the 2019 ACM SIGSAC Conference on Computer and Communications Security*, 973–990. https://doi.org/10.1145/3319535.3354212

van Leeuwen, T. (2005). Style. In *Introducing social semiotics* (pp. 139–159). Routledge.

Varis, P. (2021). Digital ethnography in the study of language, gender, and sexuality. In J. Angouri, & J. Baxter (Eds.), *The Routledge handbook of language, gender, and sexuality (pp. 164–177)*. Routledge.

Venuti, L. (1995). *The translator's invisibility: A history of translation*. Routledge Press.

Vidali, D. (2010). Millennial encounters with mainstream television news: Excess, void, and points of engagement. *Journal of Linguistic Anthropology*, *20*(2), 372–388. https://doi.org/10.1111/j.1548-1395.2010.01075.x

Vygotsky, L. S. (1981). The instrumental method in psychology. In J. V. Wertsch (Ed.), *The concept of activity in Soviet psychology* (pp. 134–143). M.E. Sharpe.

Wachter-Boettcher, S. (2017). *Technically wrong: Sexist apps, biased algorithms, and other threats of toxic tech*. WW Norton.

Walker, J. (2009). Defining the object of study. In H. Clark & D. Brody (Eds.), *Design studies: A reader* (pp. 42–48). Berg.

Weber, M. (1978). Classes, status groups and parties. In W. G. Runciman (Ed.) & E. Matthews (Trans.), *Weber: Selections in translation* (pp. 43–56). Cambridge University Press.

Weidman, A. (2010). Sound and the city: Mimicry and media in South India. *Journal of Linguistic Anthropology*, *20*(2), 294–313. https://doi.org/10.1111/j.1548-1395.2010.01071.x

Weiser, M. (1994). *Creating the invisible interface*. 7th annual ACM symposium on user interface software and technology, New York. https://doi.org/10.1145/192426.192428

Wilf, E. Y. (2014). Semiotic dimensions of creativity. *Annual Review of Anthropology*, *43*(1), 397–412. https://doi.org/10.1146/annurev-anthro-102313-030020

(2019). *Creativity on demand: The dilemmas of innovation in an accelerated age*. University of Chicago Press. https://doi.org/10.7208/chicago/9780226607023.001.0001

Winner, L. (1980). Do artifacts have politics? *Daedalus*, *109*(1), 121–136.

Woolard, K. A., & Schieffelin, B. B. (1994). Language ideology. *Annual Review of Anthropology*, *23*(1), 55–82. https://doi.org/10.1146/annurev.an.23.100194.000415

Woydack, J. (2017). Call center agents and the experience of stigma. *Working Papers in Urban Language & Literacies, Paper 215*, 2–19.

Woydack, J., & Rampton, B. (2016). Text trajectories in a multilingual call centre: The linguistic ethnography of a calling script. *Language in Society*, *45*(5), 709–732. https://doi.org/10.1017/S0047404516000610

Wysocki, A. F., & Jasken, J. I. (2004). What should be an unforgettable face. . . *Computers and Composition*, *21*(1), 29–48. https://doi.org/10.1016/j.compcom.2003.08.004

Yan, S., & Ramachandran, P. G. (2019). The current status of accessibility in mobile apps. *ACM Transactions on Accessible Computing*, *12*(1), 1–31. https://doi.org/10.1145/3300176

# Index

For EU product safety concerns, contact us at Calle de José Abascal, 56–1°,
28003 Madrid, Spain or eugpsr@cambridge.org.

www.ingramcontent.com/pod-product-compliance
Lightning Source LLC
LaVergne TN
LVHW011813200926
848484LV00005B/1181

* 9 7 8 1 0 0 9 5 4 0 5 8 2 *